高校专门用途英语（ESP）系列规划教材

U0157108

航空专业英语

Aviation English

主　编　刘泉阿荣

副主编　红　鸽

编　者　萨日娜

北京师范大学出版集团
BEIJING NORMAL UNIVERSITY PUBLISHING GROUP
安徽大学出版社

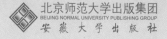

图书在版编目（CIP）数据

航空专业英语 ／ 刘泉，阿荣主编 . —合肥：安徽大学出版社，2022.4
高校专门用途英语（ESP）系列规划教材

ISBN 978-7-5664-2369-6

Ⅰ . ①航… Ⅱ . ①刘… ②阿… Ⅲ . ①航空—英语—高等学校—教材 Ⅳ . ① V2

中国版本图书馆 CIP 数据核字（2022）第 007777 号

航 空 专 业 英 语
Hangkong Zhuanye Yingyu

刘 泉 阿 荣 主编

出版发行：北京师范大学出版集团
安 徽 大 学 出 版 社
（安徽省合肥市肥西路 3 号 邮编 230039）
www.bnupg.com.cn
www.ahupress.com.cn
印　　刷：安徽利民印务有限公司
经　　销：全国新华书店
开　　本：184mm×260mm
印　　张：11
字　　数：288 千字
版　　次：2022 年 4 月第 1 版
印　　次：2022 年 4 月第 1 次印刷
定　　价：42.00 元
ISBN 978-7-5664-2369-6

策划编辑：李 梅 李 雪　　　　　　**责任印制：**赵明炎
责任编辑：李 雪　　　　　　　　　　**装帧设计：**李 雪 李 军
责任校对：高婷婷　　　　　　　　　　**美术编辑：**李 军

Preface
前 言

随着我国航空业的快速发展以及航空业国际化水平的提高，对航空相关专业学生和从业人员的航空专业英语能力要求越来越高。而航空专业英语内容多而散，专业性强，因而学习难度大。目前的航空专业英语教材，课文内容普遍为科技说明文，语言工具性较强，内容比较枯燥，远离生活实际，中国元素较少，"课程思政"元素很难挖掘。

为此我们根据中国民用航空总局颁发的有关机务维修人员和管制、签派人员的英语要求，对目前国内外的航空英语教材进行了大量的调查研究，并对航空公司、维修单位和机场的用人需求做了广泛的调研后，结合教材编写的思政纲要编写了这本《航空专业英语》。

本书以轻松明快的叙事方式对机务维修和管制（签派）的航空专业英语核心知识进行了系统化的介绍，将航空工作的责任和纪律要求以一种润物无声的方式融入了课文。同时书中还加入了我国自行研制的国产大飞机和北斗卫星导航系统的相关知识的内容，让学生了解我国的航空业发展，激发他们航空报国的使命感和责任感。

本书共有8个单元，包含飞机主要结构与部附件系统、飞行原理、飞机性能、通信导航、机场运行和交通管制、航空气象，辅以相应的音视频，使学习者能够以立体化、直观化的方式深入学习。

每个单元共有4个版块。

第1个版块为听说，围绕单元主题设计听力和会话练习，呈现相关场景中常用的短语与表达方式，培养学生的英语听说能力。内容包括专业词汇学习；经典简介/回顾，引导学生听相应内容并用自己的语言进行复述或总结；场景对话，学习在专业场景下的航空英语口语及陆空通话实际用语；相关讨论，引导学生根据话题进行英语口语交流。

第2个版块为阅读，分为精读和略读2个部分，配以相应练习，帮助学生检测学习效果。精读课文 Text A 是单元核心课文，呈现专业核心信息，侧重于语言和专业知识的输入。文本长度为1000~2000词，生词率为10%左右。单元拓展课文 Text B 是行业应用中的文本。文本长度为1000~1500词，生词率为8%左右。

第3个版块为翻译，围绕单元相关语言的功能性结构及核心要点设置翻译练习，注重培养学生对本单元主题航空英语文本的解读、转述以及实际应用能力。

第4个版块为拓展阅读，是与单元主题相关的拓展性阅读材料，以行业概述、行业故事、国内相关领域突破(如国产大飞机、北斗卫星导航系统等)为主，帮助学生开阔视野、拓宽知识面。

通过本书的学习，学生可以系统地理解航空专业领域的核心英语知识，掌握航空专业基本英语词汇、术语和常见缩略语，具备对航空专业英语文本、音视频材料的分析和解读能力，最终可以根据不同岗位的需求得体地完成相应的交际任务。本书既可以作为航空专业院校机务和交通专业学生的教材，也可以作为航空公司、维修单位和机场的培训教材使用。

本书的编写得到了中国民航大学冀晓东、李彦银、何振鹏和柳青老师的大力支持。因编写者水平有限，错误和不足之处敬请有关专家和读者批评指正。

编　者

2021年12月

★ CONTENTS ★

Unit 1 Aircraft Systems I

Lead-in

Airplanes come in a variety of shapes and sizes, but they all share the same basic components, which make them work. But what are all of these components, and what do they do? In this unit we are going to explain some aviation terms, discuss the construction of different types of airplanes, and explain how various systems in an airplane work. It is important to understand why airplanes are designed the way they are and how their systems function. That way, if a system malfunction occurs, you can potentially troubleshoot and/or fix the problem.

This unit will cover:
- The parts of an airplane
- The flight controls
- The powerplant and propeller

Part 1 Listening and Speaking

New Words & Expressions

fuselage *n.* /ˈfjuːzəlɑːʒ/		the main part of an aircraft in which passengers and goods are carried 机身
wing *n.* /wɪŋ/		one of the large flat parts that stick out from the side of a plane and help to keep it in the air when it is flying 翅膀，机翼
empennage *n.* /emˈpenɪʤ/		the rear part of an aircraft 飞机尾部
powerplant *n.* /ˈpaʊərˈplɑːnt/		also engine, a machine with moving parts that converts power into motion 发动机，动力装置
stabilizer *n.* /ˈsteɪbəlaɪzə(r)/		a device that keeps sth steady, especially one that stops an aircraft or a ship from rolling to one side 稳定装置，（飞机的）安定面，（船舶的）减摇装置
taxi *n./v.* /ˈtæksi/		(of a plane) to move slowly along the ground before taking off or after landing （起飞前或降落后在地面上）滑行
takeoff *n.* /ˈteɪkɔːf/		the beginning of a flight, when an aircraft leaves the ground 起飞
stall *v.* /stɔːl/		(of a vehicle or an engine) to stop suddenly because of a lack of power or speed; to make a vehicle or engine do this 车辆或发动机熄火，抛锚
landing gear		起落架
reciprocating engine		往复式发动机
holding point		[航] 暂留点，等待点
line up		对正中线
lift off		起飞
lose of speed		失速

I. Filling the blanks with the words from the box.

① stabilizer	② empennage	③ wings	④ powerplant
⑤ four-stroke	⑥ reciprocating engines	⑦ fuselage	

Planes come in all shapes and sizes, but they all share the same basic design components.

The _____ are shaped to maximize the amount of lift they produce, which can be attached at the top, mid-way, or at the bottom of the _____. The wings of an aircraft generate lift as air flows around them. Most planes have a single set of wings, referred to as a monoplane, but some planes have two or three sets of wings, referred to as biplanes and tri-planes, respectively.

The _____, derived from a French word having to do with "feathering an arrow", is commonly referred to as the tail section of the aircraft. It consists of two important surfaces: the horizontal and vertical _____. These stabilizers are surfaces on the tail that keep the airplane under control while flying through the air.

An airplane has an engine, which is commonly referred to as a _____. The reason behind this name is simple: the engine not only powers the airplane to move through the air, it also has other components attached to it in order to create electricity, vacuum suction, and heat, just to name. _____ have several cylinders. Inside of those cylinders, fuel and air are mixed, compressed, and then ignited. As this fuel/air mixture is ignited, its explosive force moves the piston inward. The cylinders undergo a continuous _____ cycle. The four strokes are called intake, compression, power, and exhaust.

II. Listen to the dialogue and answer the following questions.

C – Air Traffic Controller

P – Pilot

GEC – the plane that is to be taking off

P: D Ground, GEC, we are following the B737 and proceeding to the holding point.

C: GEC, you are too close to the preceding traffic. Slow down your taxi. Taxi with caution in such snowy, cold weather. GEC, reduce your taxi speed. You are maneuvering closely behind the aircraft.

P: Ground, the warmth from the B737 engines would melt the snow and ice on wings. There are not much on the wings.

C: GEC, you are forbidden to do that.

P: Tower, GEC, the aircraft is cleared of the snow, request lineup and take off.

C: GEC, are you ready for immediate departure? There is an aircraft 4 km on final.

P: GEC, affirm.

C: GEC, line up and take off immediately.

P: GEC, we are taking off now.

C: GEC, your aircraft was rolling almost 800 meters further down the runway than is customary. Do you have any problems?

P: Tower, the trip over the runway was extremely rough. The engines did not develop as much power as it needed for takeoff. We have a very difficult time in lifting off.

C: GEC, Departure, do you have any abnormal situations?

P: Departure, now the engine pressure ratio is high. The aircraft is stalling. The aircraft is falling. We are going down.

Questions

1. Why did the controller ask the pilot to slow down?

2. Why did the pilot reject the controller?

3. Why did the controller instruct the pilot to take off immediately?

4. What happened to the aircraft after takeoff, and why?

III. Look at the picture below and describe the components in detail.

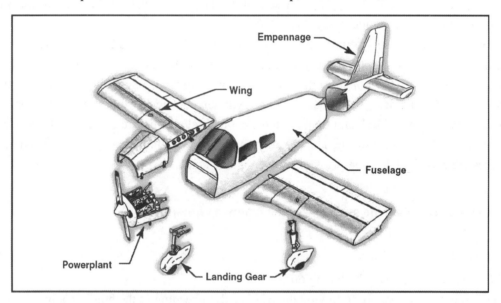

Part 2 Reading

Text A Intensive Reading

01 Parts of an Airplane

Airplanes are made up of hundreds, even thousands of parts, from the simplest pieces of wood and fabric, to newly-designed composite **airframe**s, to the most sophisticated electrical components. Planes come in all shapes and sizes, but they all share the same basic design components.

The basic components of any airplane are the fuselage, the wings, the empennage, the landing gear, and the powerplant.

The fuselage houses the cabin and **cockpit** to hold the pilots, passengers, and cargo. The fuselage is considered to be the central component of the airplane, since all the other components are attached to it. Most airplanes manufactured today use something called **semi-monocoque** construction. This means that underneath the skin of the airplane there are a series of **bulkheads** and other supports that help hold the airplane together.

The wings of an aircraft generate lift as air flows around them. The wings are shaped to maximize the amount of lift they produce. The wings can be attached at the top, mid-way, or at the bottom of the fuselage. Most planes have a single set of wings, referred to as a **monoplane**; but some planes have two or three sets of wings, referred to as **biplanes** and **triplane**s, respectively.

The empennage, derived from a French word having to do with "feathering an arrow", is commonly referred to as the tail section of the aircraft. It consists of two important surfaces: the horizontal and vertical stabilizers. These stabilizers are surfaces on the tail that keep the airplane under control while flying through the air.

Beneath the fuselage sits the landing gear, also known as the **undercarriage**. This structure is used to support the aircraft while on the ground. There are generally two different types of **configuration**s: tricycle and conventional gear. Tricycle gear is so named because its wheel configuration resembles that of a child's tricycle — that is, it has one lead wheel near the plane's nose, and two main wheels behind it, typically under the wings. Today, this is the more popular of the two types because it allows for easier landings, and improved visibility while moving on the ground, also known as taxiing. Conventional gear, on the other hand, is the older style that was popular a few decades ago. Similar to before, two main wheels are typically situated under the wings, but the third wheel is located under the airplane's tail. This results in the tail of the plane being very close to the ground, while the front of the airplane sits much higher. This obstructs the pilot's forward view because of the airplane's nose-high attitude. However, it does allow for larger engines and **propellers** to be installed on the aircraft. This setup is also the less stable of the two, making landings more difficult, especially in a **crosswind**.

Finally, there is the powerplant, which **in layman's terms** is the engine. Airplanes can have one engine, called a single-engine airplane, or have multiple engines, called multiengine airplanes. In most smaller, general aviation aircraft, **reciprocating** engines are used, like the ones found in cars. In many ways, however, airplane engines are simpler than car engines. The most basic of reciprocating engines on airplanes are not computer-controlled, are not liquid-cooled, and are not even fuel-injected.

With reciprocating engines also comes the requirement of a propeller. Just as there are different engine configurations, there are also different types of propellers. These propellers can range from the simplest two-**blade**d propeller made of a solid piece of wood, to a complicated, multi-bladed propeller with additional built-in features. These features can include: the ability to shed off ice, or even change their blade angle during flight. The rotating of the blades is similar in purpose to a car's **transmission**.

02 Flight Controls

The flight controls consist of various **surface**s around the aircraft that manipulate the **aerodynamic** forces on the plane, allowing the pilot to control the aircraft. Aircraft flight controls are broken into two systems: **primary and secondary flight controls**. Primary flight controls are simply those flight controls that the pilot primarily uses to control the airplane. The three primary flight controls are the **ailerons**, **elevator**, and **rudder**. Secondary flight controls, on the other hand, are used to change the airplane's performance and lighten the pilot's workload. The two secondary flight controls that we'll discuss are the **flaps** and **trim**.

The pilot controls the ailerons and elevator with a **yoke** or **stick**, and the rudder with the rudder **pedal**s. In most general aviation airplanes, as the pilot moves the controls, he or she is moving steel **cable**s or push rods, connected through other **linkage**s, that physically move these controls. As the control surface is deflected, the **airflow** is changed, which results in an aerodynamic force, changing the airplane's path through the air. Let's investigate the three primary flight controls in more detail.

The ailerons are located on the back end of the wings, out towards the tip, and control the aircraft's **roll** or **bank**. When the pilot moves the controls to the left, the left aileron is deflected up, creating a downward force, and the right aileron is deflected down, creating an upward force. This results in the airplane rolling to the left. The opposite would happen if the pilot were to move the controls to the right.

The elevator is attached to the back end of the horizontal stabilizer and controls the airplane's **pitch**, which allows the airplane to climb or **descend**. When the pilot moves the controls forward or **aft**, the elevator **rotate**s, **deflect**ing the air and creating a force that results in the airplane's pitch changing. If the pilot pulls back on the controls, the elevator will move up, creating a force that pushes the tail of the airplane down, thereby making the **nose** pitch up and causing the airplane to start climbing.

Finally, the rudder, which is controlled by the rudder pedals, is attached to the back end of the vertical stabilizer. As the pilot pushes on one of the rudder pedals, a cable connected to the rudder allows the rudder to move. Just like the other flight controls, as air moves around the deflected rudder, a force is applied, making the airplane **yaw**. Simply, if the pilot pushes the left rudder pedal, the nose of the airplane will **slide** to the left. The easiest way to **envision** this is to think of an airplane **suspend**ed on a **string,** above the ground. If you and your friends were to walk up to the rudder and push as hard as you can towards the right, what do you think would happen? The tail of the airplane would move to the right, and the nose would move to the left. It's that simple! We use the rudder in coordination with the ailerons to turn the airplane.

The primary flight controls then, fundamentally work the same. The pilot moves a control in the cockpit, which through cables and other linkages moves that control. As that control is moved, the airflow around it gets deflected, creating a force, and results in the plane either rolling, pitching, or yawing.

In order to help improve the performance of the airplane and make the pilot's job easier, most general aviation airplanes are equipped with flaps and trim, known as secondary flight controls. Let's examine these further.

The flaps are located on the backside of the wing, close to the fuselage. They are primarily used to help increase **lift** during **takeoff** and **landing**. The pilot controls the flaps by moving a **lever** in the airplane which either electrically moves the flaps by a motor, like on the **Cessna 172**; or the pilot manually moves the flaps using a lever, like on the **Piper Arrow**. In either case, as the pilot **extend**s the flaps, the shape of the wing changes, which increases lift. This allows the airplane to fly at slower airspeeds and make **steep**er **approach**es to landing. This is extremely beneficial during takeoff from a short runway because it allows the airplane to take off at a slower speed, meaning it will use less runway. During landing, the pilot uses flaps to allow the airplane to land at a slower airspeed, using less distance to **decelerate** and stop.

The other secondary flight control that we'll discuss is trim. The trim is used to make the pilot's job easier and allows the airplane to essentially fly itself with fewer control inputs by the pilot. The Cessna 172, like most training airplanes, has two trims: one that the pilot can directly control through cables on the elevator; and the other, a ground-adjustable **tab** on the rudder.

The elevator trim is typically controlled through a **wheel** inside the cockpit that is labeled "nose down" or "nose up." As the pilot moves this wheel, the cables will adjust the elevator trim tab, which is located on the aft or back end of the elevator. Usually during takeoff, the trim tab is in its neutral position, which means it is about **flush** with the elevator. As the pilot climbs, however, he or she may trim the airplane to help relieve control pressure and prevent pilot **fatigue**. If he or she wants to climb at a specific airspeed, then the trim can be adjusted to maintain that airspeed.

The rudder has a ground adjustable trim tab. That simply means that it can only be adjusted while on the ground, as the pilot has to manually move it. The rudder's ground-adjustable tab helps the pilot during climb because of the lift turning tendencies that the plane has while climbing.

It is obvious that the flight controls are a necessity for the pilot to fly the airplane. The pilot **manipulate**s these flight controls in order to achieve the desired performance out of the airplane. Now that we have a basic understanding of the components of the airplane and how the pilot controls different surfaces, let's go "**under the hood**" of the airplane and investigate the powerplant.

01 Parts of an Airplane

airframe *n.* /ˈeəˌfreɪm/ — the body of an aircraft, excluding its engines
机身

cockpit *n.* /ˈkɒkpɪt/ — the area in a plane, boat or racing car where the pilot or driver sits
驾驶舱，驾驶座

monocoque *n.* /ˈmɒnəˌkɒk/ — a type of aircraft fuselage, car body, etc., in which all or most of the loads are taken by the skin
硬壳式结构

semi-monocoque *n.* /ˈsemɪ ˈmɑːnəʊkɑːk/ — a type of aircraft fuselage, car body, etc., in which half or a part of the loads are taken by the skin
[航] 半硬壳式机身，半单壳式结构

bulkhead *n.* /ˈbʌlkˌhed/ — the wall that divides the inside of a ship or aeroplane into separate sections.
(船舱或机舱内的) 隔板

monoplane *n.* /ˈmɒnəpleɪn/ — an early type of plane with one set of wings
（早期的）单翼飞机

biplane *n.* /ˈbaɪpleɪn/ — an early type of plane with two sets of wings, one above the other
（早期的）双翼飞机

triplane *n.* /ˈtraɪpleɪn/ — an aeroplane having three wings arranged one above the other
三翼飞机

undercarriage *n.* /ˈʌndəkærɪdʒ/ — also landing gear, the part of an aircraft, including the wheels, that supports it when it is landing and taking off
起落架

configuration *n.* /kənˌfɪgəˈreɪʃn/ — an arrangement of the parts of sth or a group of things; the form or shape that this arrangement produces
布局，结构，构造，格局，形状

crosswind *n.* /ˈkrɒswɪnd/ — a wind that is blowing across the direction that you are moving in
侧风

reciprocate *v.* /rɪˈsɪprəkeɪt/ — to move backwards and forwards in a

straight line
沿直线往复移动

blade *n.* /bleɪd/ | one of the flat parts that turn around in an engine or on a propeller
叶片，桨叶

transmission *n.* /trænsˈmɪʃn/ | the system in a vehicle by which power is passed from the engine to the wheels
传动装置，变速器

in layman's terms | 简单地讲

02 Flight Controls

surface *n.* /ˈsɜːfɪs/ | a device that provides reactive force when in motion relative to the surrounding air; can lift or control a plane in flight
翼面

aerodynamic *adj.* /ˌeərəʊdaɪˈnæmɪk/
空气动力学的，[航] 航空动力学的

control *n.* /kənˈtrəʊl/ | the switches and buttons, etc. that you use to operate a machine or a vehicle
操纵装置，开关，按钮
primary flight control (system)
基本（主）飞行控制系统
secondary flight control (system)
辅助（副）飞行控制系统

aileron *n.* /ˈeɪlərɒn/ | (technical) a part of the wing of a plane that moves up and down to control the plane's balance
副翼

elevator *n.* /ˈelɪveɪtə(r)/ | a part in the tail of an aircraft that is moved to make it go up or down
升降舵

rudder *n.* /ˈrʌdə(r)/ | a vertical piece of metal at the back which is used to make the plane turn to the right or to the left
方向舵

flap *n.* /flæp/ | a part of the wing of an aircraft that can be moved up or down to control upward or downward movement
襟翼

trim *v. / n.* /trɪm/ — balance in flight by regulating the control surfaces, be in equilibrium during a flight
配平，配平板

yoke *n.* /jəʊk/ — an airplane control operating the elevators and ailerons
飞机操纵杆

stick *n.* /stɪk/ — also joystick, the control stick of a plane
（飞机的）操纵杆，驾驶杆

pedal *n* /ˈpedl/ — a flat bar that you push down with your foot in order to make a machine move or work
脚蹬子，踏板

cable *n.* /ˈkeɪbl/ — a kind of very strong, thick rope, made of wires twisted together
缆绳

linkage *n.* /ˈlɪŋkɪdʒ/ — a device that links two or more things
联动装置

airflow *n.* /ˈeəfləʊ/ — the flow of air around a moving aircraft or vehicle
气流

roll *n. / v.* /rəʊl/ — to turn over and over and move in a particular direction; to make a round object do this
（使）翻滚，滚动

bank *n. / v.* /bæŋk/ — to travel with one side higher than the other when turning
（飞机等拐弯时的）向内侧倾斜，倾斜飞行

pitch *n.* /pɪtʃ/ — the movement of a ship up and down in the water or of an aircraft in the air
（船在水上的）上下颠簸，纵摇，（飞机在空中的）俯仰

descend *v.* /dɪˈsend/ — to come or go down from a higher to a lower level
下来，下去，下降

aft *adv.* /ɑːft/ — in, near or towards the back of a ship or an aircraft
在 / 向船尾，在 / 向机尾

rotate *v.* /rəʊˈteɪt/ — to move or turn around a central fixed point; to make sth. do this
（使）旋转，转动

deflect *v.* /dɪˈflekt/ — to change direction or make sth change direction, especially after hitting sth.
（尤指击中某物后）偏斜，转向，使偏斜，使转向

nose *n.* /nəʊz/ — the front part of a plane, spacecraft , etc.
头部，头锥

yaw *v.* /jɔː/ — (of a ship or plane) to turn to one side, away from a straight course, in an unsteady way
偏航

slide *v.* /slaɪd/ — to move easily over a smooth or wet surface; to make sth move in this way
（使）滑行，滑动

envision *v.* /ɪnˈvɪʒn/ — to imagine what a situation will be like in the future, especially a situation you intend to work towards
展望，想象

suspend *v.* /səˈspend/ — to hang sth. from sth. else
悬，挂，吊

string *n.* /strɪŋ/ — material made of several threads twisted together, used for tying things together; a piece of string used to fasten or pull sth or keep sth in place
细绳，线，带子

lift *n.* /lɪft/ — the upward pressure of air on an aircraft when flying
提升力，升力

takeoff *n.* /ˈteɪkɔːf/ — the beginning of a flight, when an aircraft leaves the ground
起飞

landing *n.* /ˈlændɪŋ/ — an act of bringing an aircraft or spacecraft down to the ground
着陆

lever *n.* /ˈliːvə(r)/ — a handle used to operate a vehicle or piece of machinery
（车辆或机器的）操纵杆，控制杆

extend *v.* /ɪkˈstend/ — to make sth. longer or larger
使伸长，扩大，扩展

steep *adj.* /stiːp/ — making a large angle with the plane while rising or descending, sudden and

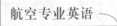

	big
	急剧（升降）的
approach *n.* /əˈprəʊtʃ/	the part of an aircraft's flight immediately before landing
	进场，进场着陆
decelerate *v.* /ˌdiːˈseləreɪt/	to reduce the speed at which a vehicle is travelling
	（使）减速行驶，降低运行速度
tab *n.* /tæb/	a short strip of material attached to or projecting from sth in order to facilitate opening or identifying or handling it
	调整片，补翼
wheel *n.* /wiːl/	a flat round part in a machine
	轮，机轮，齿轮
flush *adj.* /flʌʃ/	completely level with each other
	完全齐平
fatigue *n.* /fəˈtiːg/	a feeling of being extremely tired, usually because of hard work or exercise
	疲劳，劳累
manipulate *v.* /məˈnɪpjʊleɪt/	to control or use sth. in a skillful way
	（熟练地）操作，使用
hood *n.* /hʊd/	a cover placed over a device or machine, for example, to protect it
	（设备或机器的）防护罩，罩
under the hood	在引擎盖下面

Note

1. Cessna Aircraft Company 美国赛斯纳飞机公司

Based in Wichita, Kansas, the aviation capitol of the United States, Cessna Aircraft Company is the world's largest manufacturer of private aircraft. Cessna began its operations building small propeller-driven aircraft for the private pilot market, eventually expanding into the manufacture of corporate jets. The company has since become the leading private jet manufacturer in the industry.

The mission of the Cessna Aircraft Company is: to be the worldwide leader in the industry segments we serve by developing and producing safe, reliable, high-quality aircraft that represent the best value in general aviation; to provide the most comprehensive and responsive support to every Cessna customer; to produce the financial results that create value for Textron shareholders.

2. Cessna 172 塞斯纳 172 型飞机

The Cessna 172 is probably the World's most recognized aircraft. This amazing airplane is an American single-engine, four-seat, high wing, fixed-wing aircraft made by the legendary Cessna Aircraft Company. The Cessna 172 is the most successful aircraft in aviation history due to its popularity and longevity.

Specifications:

- Burn Rate: 8.5 gallons (32.176 litres) per hour
- Maximum cruise speed: 124 ktas (230 km/h)
- Maximum range: 640 nm (1,185 km)
- Takeoff distance: 1,630 ft (497 m)
- Ground roll: 960 ft (293 m)
- Landing distance: 1,335 ft (407 m)
- Landing ground roll: 575 ft (175 m)
- Service ceiling: 14,000 ft (4,267 m)
- Maximum climb rate: 730 fpm (223 mpm)
- Maximum limit speed: 163 kias (302 km/h)
- Stall speed: 48 kcas (89 km/h)

3. **Piper Aircraft Corporation** 美国派珀飞机公司

This company manufacturers general aviation aircraft. It was distinguished in the late 20th century as being one of the top "Big Three" in its field of general aviation manufacturing. This included Piper with other larger companies, Cessna and Beechcraft. In under 100 years of operation until 2009, Piper produced 160 different certified models of aircraft. They produce 144,000 aircraft and 90,000 of them are still in use and flying to this day.

Piper's most popular aircrafts are the Piper Cheyenne and the Piper Navajo. The Piper Cheyenne is a turboprop aircraft that has two Pratt & Whitney PT6A-28 engines. The Cheyenne can hold 4~6 passengers with a max takeoff weight of 9,000 lbs. Also, this aircraft can hit a maximum speed of 326 mph at 11,000 ft while having a range of 1,700 nautical miles. The Piper Navajo is a family class, twin-engine aircraft that are designed for the general aviation market for cargo and feeder liner operations. The Piper holds 6~8 passengers in a twin-engine corporate and commuter aircraft powered by two 310 HP Lycoming engines. Pipers CEO Simon Caldecott announced in March 2021 that he would retire and CFO John Calcagno would become the CEO of Piper. Since this change, Piper continues to be one of the world leaders in private aviation.

4. **Piper Arrow** 派珀 "弓箭手" 飞机

Piper introduced the PA-28 Cherokee in 1961 as a cheaper alternative to their PA-24 Comanche, and to compete with the Cessna 172. It was slightly smaller than the Comanche but with the same concept — a low-wing, single-engine plane perfect for use as both training aircraft and personal commuters, capable of seating four. The original PA-28 design quickly spawned many variants throughout the 1960s, including adaptation to a multi-engine. One of these variations was the Piper PA-28R Arrow.

Specifications:

- Engine: four cylinder fuel-injected Lycoming IO-360-C1C6

· Propeller: McCauley two-blade constant speed
· Wing span: 35.4 feet
· Length: 24.7 feet
· Height: 7.9 feet
· Fuel capacity: 72 US gallons
· Maximum ramp weight: 2,758 pounds
· Standard empty weight: 1,798 pounds
· Useful load: 960 pounds
· Range: 880 NM
· Service ceiling: 16,200 feet
· Takeoff distance over 50' obstacle: 1,600 feet
· Landing distance over 50' obstacle: 1,525 feet
· Cruise speed: 137 knots
· Stall speed: 55 knots, landing configuration
· Passengers: 3

Exercises

I. Answer the following questions.

1. What does the primary structure of the aircraft consist of?
2. What are the configurations of landing gear?
3. What can stabilizer do to help the airplane while flying?
4. What are the additional build-in features of propeller?
5. What are aircraft flight controls?
6. How do pilots control primary flight controls?
7. What is the main function of the secondary flight controls?
8. What is the main function of the flight controls?

II. Multiple choice (one or more answers).

1. What are the basic components of an airplane?
   A. Fuselage.
   B. Wings.
   C. Empennage.
   D. Landing gear.
   E. Powerplant.
2. The place where the pilot and flight crew, passengers and cargo sit is called _____.
   A. cabin
   B. cockpit
   C. powerplant
   D. fuselage

3. The wings can be mounted on the _____ of the fuselage.
   A. long, top and low
   B. middle, top and bottom
   C. top, tapered and middle
   D. top, middle and low
4. What is the main function of the landing gear / undercarriage?
   A. To support the aircraft while on the ground.
   B. To enable the aircraft for a taxi, safe landing and takeoff.
   C. To support the aircraft in the air.
   D. To shed off ice.
5. What is the common name for powerplant?
   A. Propeller.
   B. Transmission.
   C. Engine.
   D. Rudder.
6. Aileron is for _____ control, rudder is for _____ control and elevator is for _____ control.
   A. roll, yaw, pitch
   B. pitch, yaw, roll
   C. yaw, pitch, roll
   D. taxi, takeoff, landing
7. Where are flaps located?
   A. On the backside of the wing, far from the fuselage.
   B. On the frontside of the wing, far from the fuselage.
   C. On the backside of the wing, close to the fuselage.
   D. On the frontside of the wing, close to the fuselage.
8. Where are trims located?
   A. Elevator.
   B. Rudder.
   C. Nose.
   D. Hood.

Text B Extensive Reading

03 Powerplant and Propeller

We've just learned about the basic parts of an airplane and how it is controlled; but it's equally important for us to understand how the power is generated to move the airplane through the air. An airplane has an engine, which is commonly referred to as a powerplant. The reason behind this name is simple: the engine not only powers the airplane to move through the air, it also has other components attached to it in order to create electricity, **vacuum suction**, and heat, just to name a few. The powerplant of an airplane, like an engine of a car, is one of the most important components, because without it, there is no way to get the plane off the ground. Aviation engines can be separated into two groups: reciprocating engines and **turbine engine**s. While most airlines and **corporate airplane**s use turbine-powered airplanes, **general aviation** and **training aircraft** are equipped with reciprocating engines.

Reciprocating engines have several **cylinder**s. Inside of those cylinders, fuel and air are mixed, **compressed**, and then **ignited**. As this fuel/air mixture is ignited, its explosive force moves the **piston** inward. The pistons are connected to a **crankshaft**, and when the pistons move in and out, that causes the crankshaft to rotate. The propeller is connected to the crankshaft, so as the crankshaft rotates, so does the propeller.

The cylinders undergo a continuous four-**stroke** cycle. The four strokes are called **intake**, **compression**, **power**, and **exhaust**. The first stroke, the intake stroke, is when the piston inside the cylinder moves away from the **cylinder head**. As the piston moves away, the intake **valve** opens, and the fuel/air mixture is sucked into the cylinder's **combustion chamber**.

Once the piston has reached the base of the cylinder, it's time for the second stroke, compression. During this phase, the intake valve is closed, and the piston reverses direction, moving back towards the cylinder head. This compresses the fuel/air mixture, since it has nowhere to escape.

Once the piston approaches the top of the cylinder, we begin the third stroke: power. Two spark plugs at the head of the cylinder each lets off a **spark**, which ignites the fuel mixture and makes it combust. This controlled explosion pushes the piston back inward to the base of the cylinder, which in turn rotates the crankshaft, and, therefore, the propeller.

Finally, we reach the last stroke: exhaust. During this phase, the exhaust valve opens, and the piston moves back towards the cylinder head, pushing out the combusted gases, commonly called exhaust. Then the process starts all over again, repeating the process thousands of times every minute.

vacuum suction n. 真空吸力

turbine n. /ˈtɜːbaɪn/ 涡轮机，汽轮机

corporate airplane 公务机

general aviation 通用航空（大型营业性客运和货运之外的民航）

training aircraft 教练机，训练机

cylinder n. /ˈsɪlɪndə(r)/（发动机的）气缸

compress v. /kəmˈpres/（被）压紧，压缩

ignite v. /ɪɡˈnaɪt/（使）燃烧，着火，点燃

piston n. /ˈpɪstən/ 活塞

crankshaft n. /ˈkræŋkʃɑːft/ 曲轴，曲柄轴

stroke n. /strəʊk/ 冲程

intake n. /ˈɪnteɪk/ 进气

compression n. /kəmˈpreʃn/ 压缩

power n./v. /ˈpaʊə(r)/ 做工

exhaust n. / v. /ɪɡˈzɔːst/ 排气

cylinder head n. 汽缸盖，汽缸头

valve n. /vælv/ 阀，阀门，活门，气门

combust v. /kəmˈbʌst/ 开始燃烧，开始烧

combustion chamber n.（发动机等的）燃烧室

spark n. /spɑːk/ 火花，火星，电火花

On a typical four-cylinder engine, each one of the cylinders is in the middle of a different stroke. That way one cylinder is always in the power stroke, and the engine is able to keep the crankshaft rotating, and thereby allowing the remaining cylinders to go through their respective stroke.

As we just saw in these four strokes, there are two valves at the head of each cylinder that open and close to allow the fuel mixture in, and the exhaust gasses out. But what controls those valves? That would be the **camshaft**. The camshaft is a rotating cylinder, situated above the crankshaft, with various oblong **lobes** protruding from it. These lobes push on rods that connect to each valve, pushing them open. The valves are **spring-loaded** and will return to the closed position as the camshaft lobes move away from their respective rod.

camshaft n. /'kæmʃɑːft/ 凸轮轴

lobe n. /ləʊb/ 凸角，波瓣

spring-loaded adj. /ˌsprɪŋ'ləʊdɪd/ 弹簧承载的，弹顶的

Getting the valves to open at the exact moment is very crucial for the engine to operate. Because of that, the camshaft is **gear**ed to the crankshaft so they will remain **synchronize**d. The camshaft is geared to **spin** half as fast as the crankshaft. This results in the valves opening twice during the four-stroke cycle.

gear v. /gɪə(r)/ 使与……相适应，使适合于

synchronize v. /'sɪŋkrənaɪz/ (使)同步，在时间上一致，同速进行

spin v. /spɪn/ (使)快速旋转

Now, how do we get the fuel and air into the cylinders? It's simple: the **induction system**! Inside of the cockpit of most general aviation aircraft, there are the **throttle** and **mixture controls**. The throttle controls the amount of fuel and air that go into the cylinders, while the mixture controls how much fuel is mixed with the air. In simple terms, the mixture controls the ratio between fuel and air. Typically, for every fuel molecule, there are 15 air molecules. The mixture adjusts the amount of fuel necessary to maintain this ratio. The throttle, on the other hand, controls how much of that ratio is let into the cylinders. The more the throttle is open, the more fuel and air enter the cylinders, and therefore, the more powerful the combustion will be, making the engine run faster.

induction system n. 进气系统，感应系统，诱导体系，诱导系统

throttle n. /'θrɒtl/ 节流阀，节流杆，风门，风门杆

mixture control n. [航] 混合比调节

The air that is part of the fuel/air mixture enters the system at the **air filter**, usually found in the front of the airplane. Once the air passes through the filter, it is **meter**ed, and sent on its way to the cylinders. The fuel, on the other hand, is housed onboard the plane, typically inside of the wings. Just like the air, it is metered and then sent to the cylinders.

air filter n. /er 'miːtə(r)/ 空气滤芯，空气滤清器，空气过滤器

meter v. /'miːtə(r)/ 用仪表计量

There are two different potential systems used that can control the fuel/air mixture: the **carburetor system** and the **fuel injection system**. Most modern airplanes are equipped with fuel injection

carburetor n. /'kɑːrbəreɪtər/ 化油，化油器

systems, so we'll spend a little more time on that, but it's a good idea to still review the basics of a **carburetor**.

The job of the carburetor is to mix fuel with the air that gets sent to the cylinder combustion chambers. Fuel arrives at the carburetor and sits in the **float chamber**, waiting to be used. To the side of the float chamber is the **venturi** which is where the air passes through. As the air passes through the venturi, its **velocity** increases, which causes the pressure to decrease. Towards the bottom of the venturi, we find a fuel discharge **nozzle**, which is located near the area of low pressure. This draws the fuel out from the float chamber, through the nozzle, and mixes it with the air. Just past the venturi is the throttle valve. This controls how much of the fuel/air mixture is being sent to the cylinders.

In newer airplanes, fuel injection systems are installed, which have many benefits over **carbureted engine**s. Fuel injection engines reduce the amount of fuel required, increase engine power output, and allow for the precise use of fuel. Rather than having a carburetor, a fuel injected system is split up into different components. These consist of **fuel pump**s, a **fuel control unit**, a **fuel manifold valve**, and finally **fuel discharge nozzle**s. The fuel pump pumps fuel from the fuel tanks to the fuel control unit. Then, the fuel control unit regulates the specific amount of fuel needed based on the mixture and throttle settings. The fuel then gets sent to the fuel manifold valve where it is dispersed and heads to the nozzles of each cylinder. This time, the fuel does not mix with the air until immediately before entering the combustion chamber.

Now that we've discussed how we get the fuel and air into the engine cylinders, let's talk about how we get that mixture to ignite. The ignition system is what provides the spark to the mixture. The major components of the ignition system include the **magneto**s, the spark plugs, wires, and the ignition switch.

Just as its name implies, a magneto consists of a rotating magnet that generates sparks of electricity. The spark that is generated is independent of the airplane's electrical system. That means if the airplane's electrical system were to fail, the magnetos would still be able to generate sparks for the engine to run. The energy generated from the magnetos is sent to the spark plugs. The spark plugs then release that energy which ignites the fuel/air mixture, creating power to turn the propeller. Most airplanes have two magnetos, multiple sets of wires, and two spark plugs

carburetor system 化油器系统

fuel injection system 燃油喷射系统，燃料喷入系统

float chamber 浮箱，浮筒室，浮子

venturi n. /ven'tjʊəri/ 文丘里管（一种流体流量测定装置）

velocity n. /və'lɒsəti/（沿某一方向的）速度

nozzle n. /'nɒzl/ 管口，喷嘴

carbureted engine n. 汽化式发动机

fuel pump 燃油泵

fuel control unit 油量调节装置，燃油控制装置，燃料控制器，燃油控制组件

manifold n. /'mænɪfəʊld/ 歧管（引擎用以进气和排气）

fuel manifold valve 燃油歧管阀门

discharge n. /'dɪstʃɑːdʒ/ 排出（物）

fuel discharge nozzle 燃油排出（管嘴，接管）

magneto n. /mæg'niːtəʊ/ 磁发电机

per cylinder in order to increase efficiency and reliability of the system. If one magneto fails or one spark plug fails, the engine will still run; however, power output will be slightly reduced, as the engine is operating less efficiently. The last component is the **ignition switch**, which is the pilot's way of controlling the magnetos. In most general aviation aircraft, the ignition switch is labeled "off" "right" "left" "both", and "start". If the switch was placed "off", the magnetos would not be able to power the sparkplugs. Therefore, the engine would not be able to run and it would turn off. If the switch was placed in the "right" or "left" position, then only that respective magneto would be powering its sparkplugs. If the switch was positioned on the "both" **selector**, which is its normal position for flight, both magnetos would be powering the sparkplugs. Finally, the "start" position. This engages the **starter**, which turns the crankshaft to start the engine.

In the cylinders, the spark plugs ignite the fuel/air mixture, moving the piston in the power stroke. However, there are a few abnormal conditions that can affect the performance of the engine. The first one is called **detonation**. Detonation is the uncontrolled, explosive ignition of the fuel/air mixture inside of the cylinders. It results in high temperatures and pressures, which can cause cylinder damage. Detonation can happen for a variety of reasons including: using a lower fuel grade than what is recommended, high power settings with the mixture set too **lean**, or climbing too steeply with cylinder heads not cooling appropriately. If detonation is occurring, the pilot should reduce the power and increase the mixture, which will allow for better cylinder cooling.

Another issue that can develop is **preignition**. Just as you'd expect, preignition is simply when the fuel/air mixture is ignited prior to its normal time. This phenomenon usually occurs because of a lingering hot spot on the cylinder wall, usually caused by a **carbon deposit** that is hot enough to ignite the fuel/air mixture.

Even when the engine isn't detonating or experiencing preignition, the temperature of the engine can get quite hot. Like car engines, airplane engines have a **cooling system** in order to prevent engine damage or engine failure. There are two major ways we cool our engine: oil and air. Oil passes through an **oil cooler** and proceeds to, not only **lubricate**, but also cool the engine. As the cool oil touches the hot engine, there is a heat exchange, meaning that the oil becomes warmer, but the engine cooler. Then the oil returns back to the oil filter and oil cooler,

ignition switch 点火电门，点火开关

selector *n.* /sɪˈlektə(r)/ 选择器，转换器，换挡器

starter *n.* /ˈstɑːtə(r)/ （发动机的）启动装置，启动器

detonation *n.* /ˌdetəˈneɪʃn/ 爆震，爆震燃烧，爆炸，起爆，引爆

lean *adj.* /liːn/ （混合汽化燃料）空气占比高的

preignition *n.* /ˌpriːɪɡˈnɪʃən/ 提前点火，

carbon deposit 积炭，炭沉积

cooling system [动力] 冷却系统，散热系统

oil cooler [动力][油气] 油冷却器，油散热器

lubricate *v.* /ˈluːbrɪˌkeɪt/ 使润滑

and the process starts over. However, we also use air to cool our engines. As the airplane flies through the air, outside air flows through **inlet**s usually found on the front of the airplane like the Cessna 172. This outside air is colder than the engine, so as the air flows over the engine, it cools it in the same manner the oil cools the engine. This method of cooling can be severely restricted, however. When the airplane is running at high power and at a very low speed, such as when climbing, airflow is minimized, causing higher temperatures.

The last step in the four-stroke cycle is the exhaust. The exhaust system has a dual purpose in most general aviation airplanes. As expected, the exhaust system does allow the hot exhaust gases in the cylinder to escape the engine in a quiet manner. Additionally, as we'll talk about later, the exhaust system also provides heat to the cabin.

Now that we have a better understanding of how an engine basically works, we will examine what creates the **thrust** for the airplane to fly. Remember that as the combustion occurs in the cylinder, the pistons rotate the crankshaft, which directly drives the propeller. The propeller, when closely examined, is actually **twisted**. This is simply to create equal thrust along the propeller blade. Since the propeller rotates around its **hub**, or center, the further the propeller extends out of that hub, the faster it spins. To create equal force throughout, manufacturers twist the blade. There are two major types of propellers: **fixed pitch propeller**s and **constant speed propeller**s.

The Cessna 172 is equipped with a fixed pitch propeller. This means that the pilot cannot manipulate the propeller blade. It is installed onto the airplane, and it will always remain exactly how it was installed. An airplane equipped with a fixed propeller only has the throttle to control the engine **RPM (Revolutions per Minute)** and, therefore, propeller RPM.

A constant-speed propeller, such as the one on the Diamond DA42, is more efficient than a fixed pitch propeller. With a constant-speed propeller, the pilot can adjust the blade angle of the propeller to obtain maximum performance for different conditions of flight. Airplanes with a constant-speed propeller are not only equipped with a throttle, which still controls the power created by the engine, but they are also equipped with a propeller control. This control, which is usually blue in color, allows the pilot to change the RPM of the propeller, instead of just based on

inlet *n.* /ˈɪnlet/ （液体、空气或气体进入机器的）入口，进口

thrust *n.* /θrʌst/ （发动机推动飞机、火箭等的）推力，驱动力

twist *v.* /twɪst/ 使弯曲，使扭曲（成一定形状）
twisted *adj.* /twɪstɪd/ 扭曲的，弯曲的，变形的
hub *n.* /hʌb/ 中心，核心
fixed pitch propeller 固定螺距螺旋桨，定螺距螺旋桨
constant *adj.* /ˈkɒnstənt/ 不变的，固定的，恒定的
constant speed propeller [航]定速螺旋桨
RPM (revolutions per minute) 每分钟转数

the engine's output. The reason this propeller is called constant-speed is that when throttle movements are made, the RPM, or speed at which the propeller is rotating, will remain the same. This is accomplished by rotating the angle of the blades, which allows them to deflect different amounts of air.

Notes

Diamond DA42 奥地利钻石 DA42 飞机

The Diamond DA42 Twin Star is a four seat, twin engine, propeller-driven airplane developed and manufactured by Austrian company Diamond Aircraft Industries. It was Diamond's first twin engine design, as well as the first new European twin-engine aircraft in its category to be developed in over 25 years. In 2004, the DA42 became the first diesel-powered fixed-wing aircraft to perform a non-stop crossing of the North Atlantic.

Specifications:
- Maximum speed: 221.21 mph
- Range: 1,052 miles
- Maiden flight: Dec 09, 2002
- Length: 28 feet 1 inches
- Wingspan: 44 feet 0 inches
- Passengers: 4
- Engine: 2 Austro Engine AE300 turbocharged common-rail injected 2.0 liter diesel engine with 168 HP and EECU single lever control system
- Propeller: 2 MT propeller MTV-6-R-C-F/CF 190-69 3-blade constant speed propeller

Exercises

I. Answer the following questions.

1. What is the reason for the engine of an airplane being named as powerplant?
2. What is the four-stroke cycle that cylinders undergo?
3. How to get the valves to open at the exact moment?
4. What are the systems used that can control the fuel/air mixture?
5. What are the benefits of fuel injection systems over carbureted engines?
6. What are the major components of the ignition system?
7. How to create equal thrust through propeller?
8. Why is constant-speed propeller named so?

II. Multiple choice (one or more answers).

1. The engine can also be called _____.
   A. turbine engine
   B. powerplant

C. reciprocating

  D. turbine and reciprocating

2. The major parts of a reciprocating engine include _____.

① piston	② cylinder	③ crankshaft	④ rod	⑤ crankshaft
⑥ rod	⑦ valve	⑧ camshaft	⑨ lobe	⑩ spark plug

  A. ①②③④⑤

  B. ⑥⑦⑧⑨⑩

  C. ①③⑤⑦⑨

  D. ②④⑥⑧⑩

  E. all of them

3. The function of crankshaft is to _____.

  A. transmit the power to the propeller

  B. convert the back-forth motion

  C. produce thrust force

  D. convert mechanical energy

4. To the side of the float chamber is the _____ which is where the air passes through.

  A. venturi

  B. fuel discharge nozzle

  C. throttle valve

  D. air filter

5. What is not the major component of fuel injected system?

  A. Fuel pump.

  B. Fuel control unit.

  C. Fuel manifold valve.

  D. Fuel discharge nozzle.

  E. Carburetor.

6. If the engine has no electricity to run, which position / selector would the ignition switch be placed?

  A. Off.

  B. Right.

  C. Left.

  D. Both.

  E. Start.

7. There are a few abnormal conditions that can affect the performance of the engine. What are they?

  A. Detonation.

  B. Preignition.

  C. Overheating.

D. Electrical failure.

8. The aircraft is propelled by _____.

    A. propeller

    B. engine

    C. gas

    D. engine-propeller combination

Part 3 Translation

1. Airplanes are made up of hundreds, even thousands of parts, from the simplest pieces of wood and fabric, to newly-designed composite airframes, to the most sophisticated of electrical components.

_____

_____

2. Most airplanes manufactured today use something called semi-monocoque construction. This means that underneath the skin of the airplane there are a series of bulkheads and other supports that help hold the airplane together.

_____

_____

3. Most planes have a single set of wings, referred to as a monoplane; but some planes have two or three sets of wings, referred to as biplanes and tri-planes, respectively.

_____

_____

4. The setting of conventional landing gear obstructs the pilot's forward view because of the airplane's nose-high attitude. However, it does allow for larger engines and propellers to be installed on the aircraft. This setup is also the less stable of the two, making landings more difficult, especially in a crosswind.

_____

_____

5. The most basic of reciprocating engines on airplanes are not computer-controlled, are not liquid-cooled, and are not even fuel-injected.

_____

_____

6. As the control surface is deflected, the airflow is changed, which results in an aerodynamic force, changing the airplane's path through the air.

_____

_____

7. The elevator is attached to the back end of the horizontal stabilizer and controls the airplane's pitch, which allows the airplane to climb or descend.

8. Finally, the rudder, which is controlled by the rudder pedals, is attached to the back end of the vertical stabilizer.

9. The flaps are located on the backside of the wing, close to the fuselage. They are primarily used to help increase lift during takeoff and landing.

10. 配平板的作用是减轻飞行员的操控强度，使飞行员可以用更少的操控使飞机实现自主飞行。

11. 发动机是最重要的飞机组件之一，用来给飞机提供推力。

12. 发动机有两种：活塞式发动机和涡轮式发动机。活塞式发动机通常经历四冲程：进气、压缩、动力和排气。涡轮式发动机有几种形式：涡喷、涡桨、涡扇和涡轴。涡轮发动机组件包括进气口、压气机、燃烧室、涡轮和排气口。

13. 发动机故障有：发动机熄火、发动机失效、发动机起火、发动机揣震、发动机顺桨、发动机解体等。

14. 发动机熄火指由燃烧室里火焰熄灭导致的发动机失效。其由燃油耗尽、压气机失速、氧气供给不足、异物吸入、极端天气或机械故障导致。为了解决发动机熄火，飞行员应该重新供给燃油，重启故障发动机。

15. 发动机起火指发动机着火或者烟雾或火焰从发动机冒出。发动机起火的原因有很多：第一，易燃物质如燃油、润滑油或液压油所致；第二，异物吸入；第三，发动机机械故障；第四，维修失误。

Part 4 Supplementary Reading

C919, the First Aircraft Independently Developed by China, Commences Its Maiden Flight in Shanghai (Part 1)

The first C919 aircraft commences its maiden flight at Shanghai Pudong International Airport successfully on May 5th, 2017. The CPC Central Committee and the State Council sent congratulatory messages for the success of the first flight of C919 aircraft.

It was 15:19 on May 5th, 2017 in Shanghai, China. Shanghai Pudong International Airport, which is exceptionally busy in ordinary days, held its breath, stared soulfully, and opened its arms: An airliner with blue painting to symbolize the sky and green painting to symbolize the earth on the rear fuselage lightly spread the wings of youth and steadily landed on the fourth runway. This is a historic moment. It indicates that the "Trunk Liner Dream" haunting the Chinese nation for a hundred years has finally made a historic breakthrough, and the blue sky finally has a trunk liner which is developed by China completely in accordance with the world advanced standards. It means that after nearly half a century of hard exploration, under the leadership of the CPC Central Committee with Comrade Xi Jinping as the core, China has gained the core competences of developing a modern trunk liner. This is not only a major historic breakthrough in China's aviation industry, but also a significant achievement made by China in deeply implementing the strategy of driving development by innovation and comprehensively promoting the structural reform of the supply side.

Mr. Ma Kai, Member of the Political Bureau of the CPC Central Committee and Vice Premier of the State Council, and Mr. Han Zheng, Member of the Political Bureau of the CPC Central Committee and Secretary of the Shanghai Municipal Party Committee, attended the first flight activities. Vice Premier Ma Kai made a speech at the first flight activities. Mr. Ding Xuedong, Deputy Secretary-General of the State Council, Mr. Miao Wei, Minister and Secretary of the Party Leadership Group of Ministry of Industry and Information Technology, Mr. Xiao Yaqing, Minister of State-owned Assets Supervision and Administration Commission of the State Council (SASAC), Mr. Feng Zhenglin, Vice Minister and Deputy Secretary of the Party Leadership Group of Ministry of Transport and Director and Secretary of the Party Leadership Group of Civil Aviation Administration of China (CAAC), Mr. Zhang Yanzhong, Chairman of the Advisory Committee for the major projects of large aircraft of the State Council and Academician of the Chinese Academy of Engineering, Mr. Jin Zhuanglong, Chairman and Secretary of the Party Committee of Commercial Aircraft Corporation of China, Ltd. (COMAC) and General Director of C919 Program, and Mr. He Dongfeng, President of COMAC and Executive Deputy General Director of C919 Program, also attended the activities. Mr. Ding Xuedong read out the congratulatory messages sent by the CPC Central Committee and the State Council for the success of the first flight of C919 aircraft. Mr. Miao Wei and Mr. Feng Zhenglin made a speech at the first flight activities respectively. Mr. He Dongfeng presided over the first flight activities.

In the afternoon of May 5th, 2017, the first C919 aircraft flown by Pilot in Command Cai Jun and Test Pilot Wu Xin and carrying Observer Qian Jin and Flight Test Engineers Ma Fei and Zhang Dawei took off from the fourth runway of Shanghai Pudong International Airport at 14:00 and soared into the sky. After completing the scheduled flight test subjects with a smooth cruising flight of 1 hour and 19 minutes at an altitude of 3000 meters within specified airspace in the southeast of Nantong, the aircraft returned and landed safely at 15:19. Mr. Cai Jun reported on behalf of the crew that the aircraft functioned well in the air. Mr. Jin Zhuanglong, General Director of C919 Program, announced that the first flight of C919 was successful.

C919 aircraft is the first single-aisle trunk liner developed by China completely in accordance with the airworthiness standards and the prevailing market standards. C919 aircraft is designed to be the most popular 150-seat single-aisle aircraft in the air transport market. The baseline version has a standard range of 4,075 km and an extended range of 5,555 km, with seats for 158 passengers in hybrid configuration and seats for 168 passengers in all economy-class configuration. By using advanced aerodynamic layout, structural material and airborne system, the drag of C919 aircraft is 5% lower than that of the similar type in service, the field noise is over 10 db lower than the fourth stage requirements of International Civil Aviation Organization (ICAO), the carbon dioxide emission is 12%~15% lower, the nitrogen oxide emission is over 50% lower than the emission level specified by ICAO CAEP6, and the direct operating cost is reduced by 10%. Up to now, a total of 570 C919 orders are obtained from 23 domestic and foreign customers. C919 aircraft has strictly implemented China Civil Aviation Regulations (CCAR) Part 25 Airworthiness Standards: Transport Category Airplanes (CCAR25). CAAC accepted the application for type certificate for C919 in 2010 and comprehensively carried out airworthiness examination. European Aviation Safety Agency (EASA) accepted the application for type certificate for C919 in April 2016.

Unit 2　Aircraft Systems II

Lead-in

Aircrafts are complex products comprised of many subsystems which must meet demanding customer and operational lifecycle value requirements. It is essential to know what are all of these systems and what do they do. In this unit we are continually going to explore why airplanes are designed the way they are and how their systems function.

This unit will cover:
- The landing gear
- The fuel, oil, and hydraulic systems
- The electrical system
- And finally, the environmental system

Part 1 Listening and Speaking

New Words & Expressions

inboard *adj.* /ˈɪnbɔːd/ — located within the hull or nearest the midline of a vessel or aircraft
（飞机的）舷内的，舱内的

display *n.* /dɪˈspleɪ/ — the words, pictures, etc. shown on a screen; screen
显示，显像，显示器，屏幕

steer *v.* /stɪə(r)/ — to control the direction in which a boat, car, aircraft, etc. moves
驾驶（船、汽车、飞机等），掌控方向

squawk *n.* /skwɔːk/ — reply machine
应答机，识别报告，识别代码

QNH *n.* — altimeter sub-scale setting to obtain elevation when off the ground
高度表刻度设置以得到离地时的海拔高度

Roger *interj.* /ˈrɒdʒə(r)/ — people say "Roger!" in communication by radio to show that they have understood a message
（用于无线电通讯，表示已听懂信息）信息收到，明白

hydraulic *adj.* /haɪˈdrɔːlɪk/ — moved through pipes, etc. under pressure
（通过水管等）液压的，水力的

dispatcher *n.* /dɪˈspætʃə(r)/ — a person whose job is to see that trains, buses, planes, etc. leave on time
（火车、汽车、飞机等的）调度员

tow *v.* /təʊ/ — to pull a car or boat behind another vehicle, using a rope or chain
（用绳索）拖，拉，牵引，拽

carbon brake 碳刹车片
anti skid 防滑的
spoiler position 扰流板位置，扰流板偏度
ILS Instrument Landing System 仪表着陆系统

ILS approach 仪器进场

I. Listen to the passage and fill in the blanks with the words you hear.

The A320 is equipped with dual wheel main gear which retracts_____, and a dual wheel nose gear which retracts _____. The wheels of the main landing gear are equipped with _____ for efficient braking even at high temperature, an _____ system (A/SKID), and an _____ system (AUTO/BRK). The nose gear is equipped with _____. The ECAM (Electronic Centralized Aircraft Monitor) Wheel page displays indications for the _____, _____, _____, and _____. In the center of the ECAM wheel page, green and amber messages can be displayed to provide _____ indications. Since the Wheel page is displayed during landing, _____ is also displayed at the bottom of the page for quick reference.

Let's go to the cockpit to locate the controls and indicators for the landing gear, brakes and steering. The Landing Gear _____ is located on the Center Instrument Panel. Just above the landing gear lever is located a panel which contains switches and indicators for the Landing Gear, the Auto Brake (AUTO/BRK), the anti skid (A/SKID) and Nose Wheel Steering (N/W STRG). "Up" "Down" and "_____" indications on the gear panel are associated with position indicators on the Wheel page. When autobrakes are used, selection of Low (LO), _____ (MED) or Max (MAX) is made, using these pushbutton switches. This switch controls both the Anti-Skid and the Nose Wheel Steering through the (BSCU). Left in the "ON" position for all normal operations, it is switched off only in case of specific _____. The Landing Gear _____ handle is located on the Center Pedestal. The Steering handwheels are located on each side of the cockpit, so either pilot can taxi the aircraft. The rudder pedals can also be used to steer the aircraft. (Note: United Airlines (UAL) policy states that _____ may taxi the aircraft using the handwheel.)

II. Listen to the dialogue and mark the statements true (T) or false (F).

C – Air Traffic Controller P1 – CSN6946 Pilot   P2 – GLO386 Pilot
P1: Guangzhou Approach, CSN6946, maintain 4,200m, squawk 3151, received information K.
C: CSN6946, radar contact, GYA01 arrival, runway 021, descend 3,600 m on QNH 1022.
P1: GYA01 arrival, runway 02L, descend to 3,600m QNH 1022, CSN6946. Approach, CSN6946, we have low hydraulic warning.
C: CSN6946, roger, what is your intention?
P1: We will have a further check of the green hydraulic system. Approach, CSN6946, we tried to resume operation of the hydraulic system, but failed. We will extend the landing gear by weight. We have already telephoned out dispatcher to inform the ground mechanics to be ready for towing our aircraft.
C: Roger, any assistance do you need us to do for you?
P1: Negative, CSN6946, we will have established the landing configuration earlier.
C: Roger, CSN6946, turn left heading 050, descend to 900 m QNH, cleared ILS approach runway 02L, report established. CSN6946.
P1: CSN6946, established ILS runway 02L.
C: CSN6946, contact tower 118.8, good day.
P1: 118.8, CSN6946, good day. Guangzhou Tower, CSN6946, established ILS runway 02L.

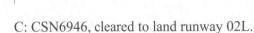

C: CSN6946, cleared to land runway 02L.

P1: Cleared to land runway 02L, CSN6946.

C: CLO386, go around immediately. Follow the standard procedure.

P2: Going around, GLO386.

C: GLO386, an airbus 320 got stuck on the runway due to hydraulic problem. Landing runway change to runway 02R, now turn left heading 270, radar vector to runway 02R.

P2: Copied that, turn right heading 270.

Statements:

1. _____ There was something wrong with the CSN6946's hydraulic system.

2. _____ The flight crew of CSN6946 wanted to do some troubleshooting.

3. _____ They solved the problem by their own.

4. _____ They failed to extend the landing gear during landing.

5. _____ Another aircraft had to go around because CSN6946 failed to vacate the runway.

III. Look at the picture and describe the components of the aircraft in the picture.

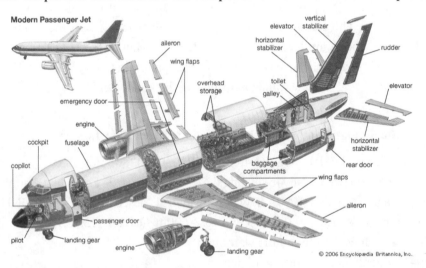

Part 2 Reading

Text A Intensive Reading

04 Landing Gear

The landing gear system provides the ground support to the airplane for taxi, takeoff, and landing. There are many types of landing gear systems, with the most common landing gear systems utilizing wheels. Some airplanes do not have wheels. What? No wheels? Yes, some airplanes are designed to operate in water, and those airplanes are equipped with **float**s.

As previously discussed, airplanes can either contain conventional or tricycle landing gear systems. Airplanes equipped with conventional gear systems are controlled on the ground through the use of **differential braking**. If a pilot wants to turn left, they step on the left brake, which slows down the left wheel, thereby rotating the plane. With tricycle landing gear systems like those the Cessna 172, as the pilot pushes on the rudder pedals, the nose wheel rotates, allowing the pilot to steer the airplane. Differential braking can also be used on tricycle gear systems to aid the pilot in making **tighter turn**s.

An airplane's landing gear system is also classified in two other categories: fixed or retractable. A **fixed landing gear system** is one in which the landing gear is permanently extended. Most primary training airplanes like the Cessna 172 are equipped with this system as they are simpler to use, there is no threat to forgetting to put the gear down for landing, and they are cheaper systems, requiring very little maintenance. The wheels of a fixed landing gear system are connected to **strut**s, which absorb the shock of landing and taxiing.

Some general aviation airplanes, like the Piper Arrow or Diamond DA42, are equipped with a **retractable landing gear system**. After takeoff, the pilot moves a handle, which electrically powers a **hydraulic pump** that allows the landing gear to rise into the body of the airplane. Prior to landing, the pilot moves the handle down, which again activates the pump, allowing the gear to come down and lock into position. The benefit of having a retractable landing gear system is that it increases performance by **streamlining** the airplane.

05 Fuel System

After learning about how the engine works, we remember that there are two things that are mixed together for combustion: fuel and air. Let's learn some things about the fuel system. The fuel system is designed to bring fuel from the **fuel tank**s to the engine. In general aviation airplanes, fuel systems can be broken down into two subsystems: **gravity-feed system**s and **fuel pump** systems.

High-wing airplanes, such as the Cessna 172, are normally designed to be gravity-feed systems. Because the fuel tank is inside the wing, which is located above the engine, gravity can naturally pull the fuel from the tank to the engine. Although, in normal operation, a fuel pump is not required, many high-wing airplanes are still equipped with some sort of fuel pump, either engine-driven, electrically powered, or both as a **backup** for emergency operations.

In **low-wing airplane**s, such as the Piper Arrow, a fuel pump is required to send fuel from the tanks to the engine. Since the fuel is located below the engine, it is simply impossible for gravity to send fuel upwards. For this reason, low-wing airplanes are designed with a fuel-pump system. Normally, there are two fuel pumps. One of the fuel pumps is engine-driven, and the backup is electrically driven.

Some airplanes are equipped with a **fuel primer**. The fuel primer allows fuel to be injected straight into the cylinder prior to engine start. Fuel priming is especially useful during cold operations. While some airplanes are equipped with a fuel primer, others, like the Cessna 172, are not. However, we'd still like to enjoy the benefits of priming the engine; therefore, it is possible to prime the engine of the Cessna 172 using the electric fuel pump.

Fuel is stored in fuel tanks, which are usually located inside of the wings. These tanks are filled from the top. There is a **cap** on the top that is easily twisted off, and gas is pumped into that tank until it is visually full. This cap is usually **vent**ed to allow outside air pressure to enter the fuel tank in order to prevent a vacuum from being created inside of the fuel tank. If the fuel tank did not have a vent, fuel would initially flow from the tank, but as it empties, there would be nothing to replace that space, which would create a vacuum and eventually stop fuel flow to the engine. Therefore, manufacturers vent the tanks to allow air to replace the fuel that is leaving the tank to the engine. Some airplanes not only have vented fuel caps, but they also have fuel vents that protrude from the wing. As the airplane flies through the air, air enters this vent and is pushed into the fuel tank.

Also inside the fuel tanks are **fuel sensor**s. These fuel sensors send information to the **fuel gauge**, which shows the pilot the amount of fuel in the tank. These sensors work in many different ways depending on the manufacturer of the airplane. It is absolutely critical to visually **verify** the amount of fuel in the tank, never entirely trust the fuel gauges after visually verifying the quantity of fuel in the tanks. It's also equally important to confirm that there are no **contaminants** in the fuel, such as water. The fuel tanks are equipped with **fuel sumps or drain**s. The pilot uses a special fuel sump cup that allows fuel to flow out of the tank and into the cup. At that point the pilot visually verifies that there are no impurities in the fuel. Most general aviation airplanes utilize 100 Low-Lead fuel which is blue in color. Therefore, not only should the pilot check for impurities, the color of the fuel should also be inspected to ensure that the proper fuel was pumped into the tank.

Another component found in the fuel system is the **fuel selector**. The fuel selector usually has four different positions, but some manufactures may have fewer positions. The four positions are: "left", "right", "both", and "off". If the fuel selector is in the "left" position, then fuel is only drawn from the left tank. Similarly, if the fuel selector is in the "right" position, then fuel is only drawn from the right tank. In normal operation, the fuel selector is in the "both" position, which allows fuel from both the left and right tanks to be sent to the engine. In the "off" position, fuel cannot pass into the engine, which stops the fuel flow to the engine. Ordinarily, the fuel selector is only placed in the off position in an emergency situation, such as an engine fire. The Cessna 172 only has three positions on its fuel selector: "left", "right", and "both". A **fuel shutoff valve** is instead used to shut off the fuel supply. This valve is located just above the fuel selector, and can be activated by pulling on the **knob**.

06 Oil System

As we've discovered, an airplane's engine cannot run without fuel. Another valuable fluid that is required for the engine to run is **oil**. We've already discussed how the oil cools the engine, but other equally important functions of oil is to lubricate the moving parts of the engine, provide a protective **coating** to prevent **corrosion**, and remove dirt and other **particle**s from the engine.

There are two types of oil systems: a **wet-sump system** and a **dry-sump system**. A wet-sump system, like that of a Cessna 172, is a system in which the oil is located in a tank at the base of the engine. This makes it an integral part of the engine. On the other hand, a dry-sump system has a separate oil tank, which separates the oil system from the engine.

Let's take a journey through the oil system of a typical general aviation airplane. As oil leaves the sump, it is **route**d through a **strainer screen**, as to not allow any solid contaminates to flow through the system and damage it. After leaving the strainer, it passes through the pump. Leaving the pump, the oil is then routed into either the **oil cooler** if it is hot and needs to be cooled, or it bypasses the cooler if it's already cool enough. Next, the oil proceeds to the **oil filter**, which removes any contaminates that made it through the strainer. Finally, the oil is sent to the engine for lubrication, cooling, and cleaning. When the oil has finished its journey, gravity pulls it back to the sump.

Since oil is so important for the engine, it's imperative to check the oil level prior to flying. The minimum and maximum amounts are specified in the airplane's flight manual. While flying, it's also important to regularly check **oil pressure** and **oil temperature**. Anything outside of the normal range, which is usually **designate**d by a green **arc**, may potentially lead to major problems, including engine failure.

New Words & Expressions

04 Landing Gear

float *n.* /fləʊt/	the structure that holds the plane on water （水上飞机的）浮筒
strut *n.* /strʌt/	a long thin piece of wood or metal used to support or make part of a vehicle or building stronger 支柱，撑杆，支杆，支撑
retractable *adj.* /rɪˈtræktəbl/	that can be moved or pulled back into the main part of sth 可缩进的，可拉回的
hydraulic *adj.* /haɪˈdrɔːlɪk/	moved through pipes, etc. under pressure （通过水管等）液压的，水力的 hydraulic pump [机] 液压泵，水力泵
streamline *v.* /ˈstriːmlaɪn/	to give sth a smooth even shape so that it can move quickly and easily through air or water 使成流线型
differential braking	差动制动

tight turn	急转弯，急弯
fixed landing gear system	固定式起落架系统
retractable landing gear system	收放式起落架系统

05 Fuel System

backup *n.* /ˈbækʌp/	extra help or support that you can get if necessary 增援，后援，后备
prime *v.* /praɪm/	fill with priming liquid 提前加注点火燃料 fuel primer 气动注油（泵）
cap *n.* /kæp/	a cover or top 帽，盖
vent *n.* /vent/	an opening that allows air, gas or liquid to pass out of or into a room, building, container, etc. （气液体的）出口，进口，漏孔
v. /vent/	to make a vent on sth; allow air, gas or liquid to pass through the vent 打孔，使气液体等从漏孔进出
sensor *n.* /ˈsensə(r)/	a device that can react to light, heat, pressure, etc. in order to make a machine, etc. do sth or show sth （探测光、热、压力等的）传感器，敏感元件，探测设备 fuel sensor 燃油（箱）传感器，燃油存量传感器
gauge *n.* /ɡeɪdʒ/	an instrument for measuring the amount or level of sth 测量仪器（或仪表），计量器 fuel gauge 燃油量表，燃油面指示器
verify *v.* /ˈverɪfaɪ/	to check that sth is true or accurate 核实，查对，核准
contaminant *n.* /kənˈtæmɪnənt/	a substance that makes sth impure 致污物，污染物
sump *n.* /sʌmp/	(also **fuel sump**) an reservoir under an engine or tank that holds the oil 集油槽，油底壳
drain *n.* /dreɪn/	a pipe through which liquid is carried

away; empty of liquid; drain the liquid from

排放管，排放

fuel drain 燃油排放，燃油排放管

knob *n.* /nɒb/   a round switch on a piece of machinery or equipment

旋钮

fuel tank   燃料箱，燃油舱，油箱

gravity-feed system   重力自动给油系统

fuel pump   燃油泵

high-wing airplane   上单翼飞机

low-wing airplane   下单翼飞机

fuel selector / fuel tank selector

燃油箱选择器

fuel shutoff valve   断油阀

06 Oil System

oil *n.* /ɔɪl/   a form of petroleum that is used as fuel and to make parts of machines move smoothly

燃油，机油，润滑油

coating *n.* /ˈkəʊtɪŋ/   a thin layer of a substance covering a surface

（薄的）覆盖层，涂层

corrosion *n.* /kəˈrəʊʒən/   the damage that is caused when something is corroded, a state of deterioration in metals caused by oxidation or chemical action

腐蚀

particle *n.* /ˈpɑːtɪkl/   a very small piece of sth

颗粒，微粒

route *v.* /raʊt/   to send sb/sth by a particular route

按某路线发送

strainer *n.* /ˈstreɪnə(r)/   a tool with a lot of small holes in it, used for separating solids from liquids

过滤器，滤盆，滤网

oil strainer 油滤

screen *n.* /skriːn/   a strainer for separating lumps from powdered material or grading particles

	滤网
	strainer screen 过滤器滤网
designate *v.* /ˈdezɪɡneɪt/	to show sth using a particular mark or sign 标明，标示，指明
arc *n.* /ɑːk/	a curved shape 弧形
wet sump	湿机匣，湿式油底壳，湿滑油槽
dry sump	干机匣，干式油底壳，干滑油槽
oil cooler	[动力][油气] 油冷却器，油散热器
oil filter	滤油器，机油滤清器
oil pressure	油压
oil temperature	油温

Notes

100 Low-Lead fuel 100 低铅燃料

Avgas (Aviation Gasoline) 100LL is one of four grades of avgas and is used primarily as fuel for piston-powered craft due to its low flashpoint. Avgas 100LL is a high-octane gasoline which allows a powerful piston engine to burn its fuel efficiently, a quality called "anti-knock" because the engine does not misfire, or "knock". The suffix LL (which stands for Low Lead) describes a grade containing lower tetraethyl lead than a second grade of identical lean and rich mixture ratings. Avgas 100LL is a blue liquid with a specific gravity of 0.68 ~ 0.74 at 60ºF (15.6ºC).

Exercises

I. Answer the following questions.

1. What is the main function of landing gear?
2. What is the advantage of the retractable landing gear system?
3. Why are some high-wing airplanes equipped with fuel pump, even if they already have gravity-fed systems?
4. What is the function of vent on the fuel cap?
5. How is fuel delivered to the engines?
6. Why do some airplanes have fuel vents that protrude from the wing?

II. Multiple choice (one or more answers).

1. What are the advantages of the fixed landing gear system?
    A. Simple to use.
    B. No threat to forgetting to put the gear down for land.
    C. Cheap system.
    D. Little maintenance.

2. To which component of fuel system does fuel sensor send information?

    A. Fuel tank.

    B. Fuel sump.

    C. Fuel gauge.

    D. Fuel selector.

3. How to verify the amount of fuel in the tank?

    A. To monitor the fuel gauge.

    B. To visually check the tanks.

    C. To check the fuel flowing out of the fuel drains.

    D. To run the engine until it stops.

4. In an emergency situation, such as engine fire, what position should fuel selector be in?

    A. Left.

    B. Right.

    C. Both.

    D. Off.

5. What is the function of engine oil?

    A. To cool the engine.

    B. To lubricate the moving parts of the engine.

    C. To provide protective coating to engine.

    D. To run the engine.

6. What are the correct steps of oil travelling through the oil system?

> ① to go to the engine ② to leave the sump ③ to route into the oil cooler or bypass it
> ④ to go through strainer ⑤ to go through oil filter ⑥ to pass through pump
> ⑦ to go back to sump

    A. ②④①⑦⑤⑥③

    B. ①③⑤⑦④②⑥

    C. ②④⑥③⑤①⑦

    D. ⑥②④⑤③⑦①

Text B Extensive Reading

07 Hydraulic System

Nearly all airplanes are equipped with a hydraulic system. While small general aviation airplanes may have small, simple hydraulic systems, bigger jets have very complex hydraulic systems. In smaller airplanes, the hydraulic system powers the brakes to stop the airplane, extend and retract the landing gear, and change the blade angle on some constant-speed propellers, as previously discussed. On larger airplanes, like those used by the airlines, the hydraulic system powers a majority of the airplane including the flight controls and flaps.

Usually, a hydraulic system consists of a **reservoir** where the **hydraulic fluid** is stored, a pump that moves the fluid, a filter to keep contaminants out of the system, a **relief valve** in case of a hydraulic malfunction, and **actuators**, which the hydraulic system operates. The hydraulic system works by pumping **incompressible** fluid through **hydraulic line**s from one actuator into another, causing the **actuator piston**s to extend or contract. The hydraulic pressure exerted throughout the actuators is significant, making hydraulic systems very powerful.

Let's examine the brake system on a typical general aviation airplane. As the pilot presses on the brakes, a piston drives fluid from the brake actuator on the pedal, through hydraulic lines, and then to the actuator near the wheels. The fluid pushes a piston which then mechanically squeezes the brake pads against the brake disk, causing the airplane to slow down.

08 Electrical System

The electrical system provides electrical power to many different systems. From some flight instruments, to aircraft lights, to flaps and landing gear, the electrical system powers them all. The electrical system is usually comprised of a 14 volt or 28 volt **direct current** or DC system, made up of some basic components. These consist of: an **alternator**, a **battery**, **switch**es, **circuit breaker**s or **fuse**s, **relay**s, a **voltage regulator**, an **ammeter or load meter**, and the electrical wiring which connects it all. Let's study some of these components in more detail.

Similar to a car, the alternator is driven by the engine through an **alternator belt**, which generates electricity for the entire system. The alternator is the primary means of powering the electrical system during normal operations with the engine running. Not only does the alternator provide power to the electrical system, it also **charge**s the battery. The battery is mainly used to start the engine, or to power the equipment while the engine is not running. In cold weather, the battery capacity can be severely reduced, so it is important that the pilot conserves battery power.

The electrical system is protected either by circuit breakers or fuses. Most modern airplanes are equipped with circuit breakers

reservoir *n.* /ˈrezəvwɑ:(r)/ 储液器，液压油箱

hydraulic fluid 液压油，工级体

relief valve 安全阀，减压阀

actuator *n.* /ˈæktʃʊeɪtər/ 传动装置，促动器

incompressible *adj.* /ˌɪnkəmˈpresəbl/ 不能压缩的

line *n.* /laɪn/ 管道，线路

hydraulic line 液压管路，液压管

actuator piston 促动器活塞

current *n.* /ˈkʌrənt/ 电流

direct current (DC) 直流电

alternator *n.* /ˈɔ:ltəˌneɪtə/ 交流发电机

battery *n.* /ˈbætəri/ 电池

switch *n.* /swɪtʃ/ （电路的）开关，闸，转换器

circuit *n.* /ˈsɜ:kɪt/ 电路，线路

breaker *n.* /ˈbreɪkə/ [电]断路器，熔断机制

fuse *n.* /fju:z/ 保险丝，熔断器

relay *n.* /ˈri:leɪ/ 中继设备

voltage *n.* /ˈvəʊltɪdʒ/ 电压，伏特数

instead of fuses because they are **resettable**. A circuit breaker **pop**s when there is excessive voltage, which results in high heat in the electrical wire. Fuses are not the popular choice today. Although they work similar to a circuit breaker, once a fuse burns, it must be replaced and that electrical circuit will not be complete until that fuse is fixed.

Circuit breakers are usually grouped together in the cockpit by the **bus** they are located on. Not the bus that brings you to school, but an electrical bus! An electrical bus is like a **power strip**. You can plug multiple things into a strip, and if you turn that strip off, everything plugged into that strip will not be powered. In an electrical system, the same remains true. The components of an electrical system are usually divided into multiple buses. For example, there may be a **main bus** and an **avionics bus**. On the main bus, main equipment such as lights are powered, and on the avionics bus, most of the instruments and equipment to run those instruments are powered.

While there may be one electrical bus or many, the entire electrical system needs to be powered safely, in order to prevent a possible electrical fire. In order to do this, a voltage regulator and alternator control unit work to monitor and control the electrical system. The voltage regulator will allow the alternator's generated power to charge the battery and power the system with an acceptable voltage by stabilizing the output of the alternator.

Finally, the pilot can monitor the electrical system by either an ammeter or a load meter. An ammeter shows the performance of the electrical system. If the alternator is providing sufficient power to the electrical system and is charging the battery, then the ammeter will show a charge. If the alternator is not charging the battery, or the battery is being used because the alternator has failed, a **negative indication** can be seen on the ammeter. A zero indication simply means that it is neither charging nor discharging the battery.

While an ammeter shows electrical performance of the alternator in relation to the battery, a load meter shows the load that is being placed on the alternator. So, if the alternator's load is 40 **amp**s, then the indication will show 40. If the load meter shows zero, then the alternator is either off or has failed. If the alternator fails in any electrical system, the battery will provide power, but not for very long. Depending on the airplane and if there are any back up batteries, an alternator failure may require a diversion from the planned flight.

voltage regulator 稳压器，[电]调压器

ammeter n. /ˈæmiːtə(r)/ 安培计，电流表

load n. /ləʊd/ 供电量

meter n. /ˈmiːtə(r)/ 计量器，计量表

load meter 载荷计，载荷测定仪

alternator belt 发电机皮带

charge n. /tʃɑːdʒ/（电池或带电物质的）充电量，电荷 v. /tʃɑːdʒ/ 给……充电

resettable adj. 可复零的，可重调的

pop v. /pɒp/（使）爆裂，发爆裂声

bus n. /bʌs/ 母线，总线，信息转移通路，汇流排

strip n. /strɪp/（纸、金属、织物等）条，带

power strip 拖线板，电源板，电源插座，接线板

main bus 主母线，系统总线，主总线

avionics bus 航空电子总线

negative indication 负号指示，负指示

amp n. /æmp/ 安，安培(电流单位)

09 Environmental System

The environmental system of the airplane provides fresh air and heat to the cabin. Because airplanes operate in such different temperatures based on where they are in the world, the pilot must be able to control the airflow and heat entering the cabin in order to ensure that everyone is comfortable. In general aviation airplanes there are usually two controls: a **cabin heat control** and a **cabin air control**.

In the winter months, when the outside air temperature is colder than desired, the pilot may turn on the cabin heat. The cabin heat valve allows outside air to pass over the **exhaust muffler shroud / shield**, which heats the air. This heated air is then **duct**ed into the cabin.

During the summer, the cabin can be cooled by venting outside air in. By opening the cabin air control, outside air is **scoop**ed into the plane and vented into the cabin through the same vents as the heater. Additional vents may also be available that solely bring in cooler outside air. Many general aviation aircraft are not equipped with air conditioners. To cool off, all you have to do is simply climb to a higher altitude, where the air is cooler.

cabin heat control 座舱加热控制

cabin air control 座舱空气调节 / 控制

muffler *n.* /'mʌflə/ 排气消声器，排气管减声器

shroud *n.* /ʃraʊd/ 覆盖物，遮蔽物

shield *n.* /ʃiːld/ 护罩，防护屏，挡板

exhaust muffler shroud / shield 排气消声器罩，排气管减声器罩

duct /dʌkt/ *n.* 管道，管子 *v.* 传送

scoop /skuːp/ *n.* 勺，铲子 *v.* 像用勺子一样舀

Conclusion

Now that you have taken a look inside a typical airplane, you are well on your way to becoming more familiar with it. Just like the driver of a car, or a captain of a ship, knowing how and why things work in your aircraft will make you a safer and more qualified aviator.

Exercises

I. Answer the following questions.

1. Why is hydraulic system so powerful?
2. How does a pilot stop the airplane through hydraulic system?
3. What is the advantage of circuit breaker over fuse?
4. What do ammeter and load meter do to the electrical system?
5. What does the environmental system do to plane?
6. How does general aviation airplane cool or warm its cabin?

II. Multiple choice (one or more answers).

1. Why do smaller airplanes need to have hydraulic systems installed?

A. To power the brakes.

B. To extend landing gear.

C. To retract landing gear.

D. To change the blade angle on constant-speed propellers.

2. Where is hydraulic fluid stored?

A. Reservoir.

B. Filter.

C. Relief valve.

D. Actuator.

3. Which of the following is not the component of electrical system?

A. Alternator.

B. Circuit breaker.

C. Voltage regulator.

D. Solar panel.

4. What components ensure the entire electrical system to be powered safely?

A. Voltage regulator and alternator control unit.

B. Ammeter.

C. Load meter.

D. Avionics bus.

Part 3 Translation

1. Airplanes equipped with conventional gear systems are controlled on the ground through the use of differential braking.

_____

2. After takeoff, the pilot moves a handle, which electrically powers a hydraulic pump that allows the landing gear to rise into the body of the airplane.

_____

3. It's also equally important to confirm that there are no contaminants in the fuel, such as water.

_____

4. Fuel is first supplied to both engines from the center tank and then from the respective tanks to engines.

_____

5. There are two types of oil systems: a wet-sump system and a dry-sump system.

6. Since oil is so important for the engine, it's imperative to check the oil level prior to flying. The minimum and maximum amounts are specified in the airplane's flight manual.

7. The hydraulic reservoir is a tank or a container to store sufficient hydraulic fluid for all conditions of operation.

8. Hydraulic power is transmitted by movement of fluid by a pump. The pump does not create the pressure, but the pressure is produced when the flow of fluid is restricted.

9. 电力系统用于控制、操作和指示各种飞机系统，无论飞机是在地面还是在飞行。

10. 电源分配系统提供并控制不同飞机系统所使用的交流、直流电源。

11. 在通用航空飞机上通常有两种控制系统：座舱加热控制和座舱空气控制系统。

12. 许多通用航空飞机没有配备空调。要想凉快下来，你只需要爬到海拔更高的地方，那里的空气更凉爽。

Part 4 Supplementary Reading

C919, the First Aircraft Independently Developed by China, Commences Its Maiden Flight in Shanghai (Part 2)

The successful first flight of C919 aircraft means that China has made clustered breakthroughs in civil aircraft technology and gained the core competences of developing trunk liners. Since the research and development of C919 aircraft was started in July 2008, COMAC has taken the development path of "China design, system integration, international tendering, and gradual promotion of domestication", adhered to the technical route of "Independent development, international cooperation, and international standard", captured more than 100 key and core technologies including aircraft and engine integration, fly-by-wire system control law, active control technology and whole-aircraft refined FEA, established China civil aircraft R&D core competences including design and development, final assembly and manufacture, customer service, airworthiness certification, supplier management and marketing by taking COMAC as a platform, formed a civil aircraft industry chain leaded by Shanghai and participated by nearly 200,000 people from more than 200 enterprises in 22 provinces and cities such as Shaanxi, Sichuan, Jiangxi, Liaoning and Jiangsu, and improved China's capability to support the civil aircraft industry. COMAC has also promoted 16 international aviation enterprises to found 16 joint ventures with Chinese enterprises, which drives the industrial development of airborne systems including power unit, avionics, flight control, power supply, fuel and landing gear. 16 material manufacturers and 54 standard part manufacturers including China Baowu Iron and Steel Group Co., Ltd. have become the suppliers or potential suppliers of C919 program. A group of aviation industry supporting areas were set up in Shaanxi, Jiangsu, Hunan, Jiangxi and other provinces. A relatively complete industry chain that has innovation capability and independent intellectual property rights, centers on COMAC, unifies AVIC, radiates throughout China and faces to the whole world is under construction.

The successful first flight of C919 aircraft also means that China has made new significant achievements in implementing the strategy of driving development by innovation. Trunk liner is called "a flower in modern industry". With the delivery of ARJ21 aircraft for operation, the first flight test of C919 aircraft and the setup of Sino-Russian long-range widebody aircraft program, China civil aircraft industry is advancing rapidly towards marketization, industrialization and internationalization. Through the research and development of C919 and ARJ21 aircraft, China has mastered more than 6,000 civil aircraft technologies in 20 specialties and 5 categories, accelerated the group breakthroughs in key technologies such new material, modern manufacturing and advanced power, and promoted the development of many basic disciplines such as fluid mechanics, solid mechanics and computational mathematics. A large quantity of advanced materials represented by the third-generation aluminium-lithium alloys and composite materials are applied in the China-made civil aircraft for the first time, accounting for 26.2% of the weight of the aircraft structure,which promotes the establishment of manufacturing and process system for special materials such as landing gear 300M steel, and facilitates the application of "green" advanced processing methods such as titanium alloy 3D printing and skin mirror milling. 36 domestic colleges and universities including Tsinghua University, Shanghai Jiao Tong University, Beihang University and Northwestern Polytechnical University have participated in the technology research and development of C919 aircraft, a civil aircraft technology innovation system which gives priority to COMAC, takes market as guidance and integrates production, study and research as one was set up, and a path for the innovation and development of major national science and technology projects was preliminarily formed.

During the course of developing ARJ21 and C919 aircraft, our country has cultivated a backbone team for the development of civil aircraft with youth as the main body. Since its foundation in 2008, COMAC has insisted on the strategy of "Developing programs via talents and cultivating talents via programs". The total number of COMAC employees has increased from more than 3,000 when the company was established to more than 10,000, and a group of civil aircraft leading talents and backbones were cultivated in the aircraft research and development practice with the post-80s and post-90s as the main body. Among all the employees, and more than 160 experts are listed in various national and local talents programs. During the process of tackling key research topics and overcoming difficult problems for more than ten years, COMAC has borne in mind the spirit of "two bombs and one satellite", the spirit of manned spaceship, the spirit of servicing the nation with aviation and the spirit of craftsmen, adhered to the spirit of "settling down to, concentrating on and being infatuated with one's job", persisted in "struggling hard, tackling key problems, enduring hardship and devoting for a long term", insisted on "mastery design, fine manufacture, service in good faith and constant perfection seeking", comprehensively carried out innovation, pioneering and creation practices in program research, program development, corporate governance and party building, and fostered the spirit of Chinese trunk liner, which gave wings to the dream of letting Chinese trunk liner fly in the blue sky.

The maiden flight of C919 aircraft indicates that the program comprehensively enters the phases of development flight test and verification flight test. A total of 6 test aircraft of C919 development batch are put into flight test to comprehensively carry out the flight test related to stall, power, performance, control stability, flight control, icing, high temperature and extreme cold. Meanwhile, two aircraft for ground tests are respectively put into test such as static test and fatigue test. The second flight test aircraft is scheduled to complete the first flight in the second half of the year.

Unit 3 Principles of Flight

Lead-in

Lift, weight, thrust and drag are forces that act upon all aircraft in flight. Understanding how these force work and knowing how to control them with the use of power and flight controls are essential to flight. This chapter examines the aerodynamic of flight and the fundamental physical laws governing the forces acting on an aircraft in flight, and what effect these natural laws and forces have on the performance characteristics of aircraft.

This unit will cover:
· Lift, weight, thrust and drag
· Newton's Three Laws of Motion
· Bernoulli's Principle
· Ground effect

Part 1 Listening and Speaking

New Words & Expressions

perform *v.* /pərˈfɔːrm/	do, carry out 执行
approach *n.* /əˈprəʊtʃ/	get near 近进
descend *v.* /dɪˈsend/	go down 下降
crew *n.* /kruː/	whole staff on the plane 机组
troubleshoot *v.* /ˈtrʌblʃuːt/	solve the trouble 排除故障
runway *n.* /ˈrʌnweɪ/	a long strip on the ground for a plane to take off or land on 跑道

I. Listen to the passage and fill in the blanks with the words you hear.

On August 21, 2021, An Uzbekistan Airways Boeing 787-8, _____ flight 603 from Tashkent (Uzbekistan) to Moscow Domodedovo (Russia), was on _____ to Moscow _____ through 4,000 feet when the _____ reported problems with the _____, climbed back to 4,000 and 5,000 feet and entered a hold to _____ the problem. The aircraft subsequently landed safely on Domodedovo's _____ 32L at a higher than normal speed (about 180 knots over ground) about 25 minutes later.

II. Filling the blanks with the words from the box.

① Reaction	② deflecting	③ rotating
④ applied	⑤ downward	⑥ flows

While all Three Laws of Motion _____ to flight, the Third Law has the most significance to lift production. Newton's Third Law states that for every action there is an equal and opposite _____. If you stick your hand out of a moving car's window, you will notice that your hand will want to lift up. By _____ your hand, you were _____ the air that comes in contact with your hand downward, and as a result, the air will push your hand up. This is similar to how a wing works. In normal flight, as air _____ around the wing, the air gets deflected _____ as it flows smoothly around it, and as a result the wind will lift the wing up.

III. Read the dialogue and Discuss.

C – Air Traffic Controller

P – Pilots

P: Fastjet 372, we are in a dive here.

C: Fastjet 372, say again.

P: Fastjet 372, we are out of 26 thousand feel in a dive. We've lost control of our airplane.

C: Fastjet 372, roger.

P: We are at 270. OK, now we have it back under control here.

C: Fastjet 372, say altitude you'd like to maintain.

P: Fastjet 372, we are at 24 thousand feet, kind of stabilized. We are gonna do a little troubleshooting. Can you give me a block altitude between 20 and 25?

C: Fastjet 372, maintain block altitude FL200 to FL250.

P: Roger, thank you.

C: Fastjet 372, do you have it under control?

P: We've got some control of the plane by thrust, but no control of the elevator. Request to proceed to Sandy Island for landing.

C: Fastjet 372, fly heading 320, descend at your discretion. Emergency services will be standing by at Sandy Island.

Discussion

1. What's the function of flaps?
2. What will you do if you encounter flap problem during approach?
3. What do we know from the dialogue?
4. What should the pilot do when encountering lose of control?

Part 2 Reading

Text A Intensive Reading

Principles of Flight (1)

While an airplane is in flight, there are four forces acting upon it. They are lift, weight, **thrust** and **drag**. Lift is the upward force created by the **wing**s as air flows around them and keeps the airplane in the air. Weight is the **downward** force toward the center of the earth opposite lift, which exists due to **gravity**. Next, we have thrust. This is the forward force generally created by the **aircraft**'s **propeller**s or **turbine engine**s, which pulls or pushes the aircraft through the air. Finally, there is drag. Drag is the force acting in the direction opposite of thrust, which **fundamentally** limits the **performance** of the airplane. When an aircraft is **maintaining** its **heading**, **altitude** and **airspeed**, it is said to be in straight and **level unaccelerated flight**. In unaccelerated flight, lift equals weight and thrust equals drag. Let's look at these four forces in a little more detail.

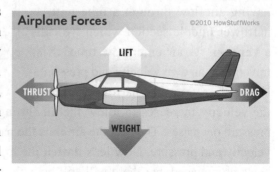

Figure 3.1 Airplane Forces

The key to an aircraft being able to fly is lift. Looking at a **cross-section** of a wing, we can better understand how the lift gets **generate**d. A wing is a type of **airfoil**. Air foils in general are just any surface that generates an **aerodynamic** force as a fluid, in our case, air moves around it. Don't **confuse** a **fluid** with a liquid. Fluids are any **substance** that **deform**s under an **applied** stress. Liquids, gases and **plasma**s are all considered fluids. In addition to the wings, all the flight control services as well as the propeller is considered airfoils. The aircraft's fuselage is even an airfoil, but it's not very good at producing lift.

Before we get to in depth, let's introduce a few new terms. The forward most point of the wing is called the **leading edge**. The aft most point is called the **trailing edge**. If we connect these two edges together with an **imaginary** line, this line is called the **chord line**. As an airplane flies through the air, the path that the plane travels along is known as its flight path. The airflow that flows around the airplane as it travels through the air is known as the relative wind. The **relative wind** is **parallel** to but opposite the aircraft's flight path, the angle between the wings cord line and the aircraft's relative wind is called the aircraft's angle of attack. The **angle of attack** is a major factor as to how much lift the wings generate. So now that we've got those terms out of the way, how does a wing actually create lift? Well, there are two major **theories** working in **unison** that explain the creation of lift. These are Newton's Three Laws of Motion and Bernoulli's Principle.

While all Three Laws of Motion applied to flight, the Third Law has the most **significance** to lift production. Newton's Third Law states that for every action there is an equal and opposite **reaction**. If you stick your hand out of a moving car's window, you will notice that your hand will want to lift up. By **rotating** your hand, you were **deflect**ing the air that comes in **contact** with your hand downward, and as a result, the air will push your hand up. This is similar to how a wing works. In normal flight, as air flows around the wing, the air gets deflected downward as it flows smoothly around it, and as a result the wind will lift the wing up.

The other main theory of lift is Bernoulli's Principle. The Principle states as the **velocity** of a fluid, in this case–air–increases, its **internal** pressure decreases. We can **visualize** this by having air flow through a tube with a narrower middle section which we call a **venturi**. As air enters the tube, it is traveling at an own velocity and pressure. When it arrives at the narrower portion,

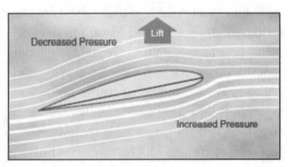

Figure 3.2 How Lift is Produced

the velocity increases to allow the air through. As the air's velocity increases, the airs pressure decreases. Then, as the air exits the narrower portion, it returns back to its **original** velocity and pressure. Now let's **flatten** the top part of the tube, **granted** the effect will not be as pronounced, but there will still be a change in velocity and pressure as the air moves through. Now how does this **relate to** a wing? Well, if we replace the bottom **protrusion** of the tube with a wing, in essence we have the same thing as a venturi. As the air passes over the wing, each **layer** of air gets deflected less and less until finally we reach a layer where the air is not disturbed at all from the wing. This can be thought of as the top of the venturi. An airplane's wing is shaped similar to that of a venturi, the top is rounded while the bottom is relatively flat. Because of this, the air traveling over the wing will increase in speed, and as a result will have a lower pressure than the air below the wing. This **imbalance** in pressure is called a **pressure gradient**. Wings are designed to create this kind of pressure gradient, because air always moves from areas of high pressure to areas of low pressure. Since the wing is stuck in between the two areas of unequal pressure, it is lifted towards the area of low pressure by the force of the higher pressure air trying to move to the low pressure side of the wing.

New Words & Expressions

thrust *n.* /θrʌst/ — the force used in pushing "the thrust of the jet engines"
推力

drag *n.* /dræg/ — aerodynamics in opposition to the movement of the aircraft, the force that impedes the advance of an aircraft
阻力

wing *n.* /wɪŋ/ — one of the large flat parts that stick out from the side of a plane and help to keep it in the air when it is flying
机翼

downward *adj.* /ˈdaʊnwəd/ — moving or pointing towards a lower level
向下的

gravity *n.* /ˈgrævəti/ — the force that on the earth pulls them towards the centre of the planet
地心引力

aircraft *n.* /ˈeəkrɑːft/ — any vehicle that can fly and carry goods or passengers
飞机

propeller *n.* /prəˈpelə(r)/ — a device with two or more blades that turn quickly and cause an aircraft to move forward
螺旋桨（飞机的推进器）

turbine *n.* /ˈtɜːbaɪn/ — a machine or an engine that receives its power from a wheel that is turned by the pressure of water, air or gas

涡轮机

engine *n.* /ˈendʒɪn/ | the part of a vehicle that produces power to make the vehicle move

发动机，引擎

fundamentally *adv.* /ˌfʌndəˈmentəli/

in every way that is important

根本上

performance *n.* /pəˈfɔːməns/ | how well or badly sth works

性能

maintaining *v.* /meɪnˈteɪnɪŋ/ | maintain something at a particular rate or level / to make sth continue at the same level,standard,etc.

保持，维持

heading *n.* /ˈhedɪŋ/ | the direction an airplane is pointed referenced to true north or magnetic north

航向

altitude *n.* /ˈæltɪtjuːd/ | the height above sea level

高度

airspeed *n.* /ˈeəspiːd/ | the speed of the airplane through the air

空速

level *adj.* /ˈlevl/ | horizontal

水平的

unaccelerated *adj.* /ˌunækˈseləreɪtɪd/

未加速的

flight *n.* /flaɪt/ | the act of flying

飞行

cross-section *n.* /ˈkrɒs ˈsekʃn/ | what you would see if you could cut straight through the middle of it

横截面，剖面

generate *v.* /ˈdʒenəreɪt/ | to produce or create sth

产生

airfoil *n.* /ˈeəfɔɪl/ | a surface designed to obtain a useful reaction,such as lift, from the air through which it moves

翼面，机翼，翼型，翼剖面

confuse *v.* /kənˈfjuːz/ | to think wrongly that sth is sth else

混为一谈，（将……）混淆

fluid *n.* /ˈfluːɪd/ | a substance that can flow

流体

substance *n.* /ˈsʌbstəns/ — a type of solid, liquid or gas that has particular qualities
物质

deform *v.* /dɪˈfɔːm/ — to change or spoil the usual or natural shape of sth
改变……的外形

apply *v.* /əˈplaɪ/ — to use sth or make sth work in a particular situation
施加

plasma *n.* /ˈplæzmə/ — a gas that contains approximately equal numbrs of positive and negative electric and is present in the sun and most stars
等离子体

imaginary *adj.* /ɪˈmædʒɪneri/ — existing only in your mind or imagination
假想的

parallel *adj.* /ˈpærəlel/ — two or more lines that are parallel to each other are the same distance apart at every point
平行的

theory *n.* /ˈθɪəri/ — the principles on which a particular subject is based
理论

unison *n.* /ˈjuːnɪsn/ — at the same time
一致

significance *n.* /sɪgˈnɪfɪkəns/ — the importance of sth, especially when this has an effect on what on what happens in the future
意义

reaction *n.* /riˈækʃn/ — a force shown by sth in response to another force, which is of equal strength and acts in the opposite direction
反作用力

rotate *v.* /rəʊˈteɪt/ — to move or turn around a central fixed point
（使）旋转

deflect *v.* /dɪˈflekt/ — to change direction or make sth change direction, especially after hitting sth
使转向，使偏斜

contact *n.* /ˈkɒntækt/ — the state of touching sth
接触

velocity *n.* /vəˈlɒsəti/ — the speed of sth in a particular direction

速度

internal *adj.* /ɪnˈtɜːnl/   connected with the inside of sth
内部的

visualize *v.* /ˈvɪʒʊəlaɪz/   to form a picture of sth in your mind
想象

venturi *n.* /venˈtjʊəri/   a part of pneumatic duct with reduced flow area for airflow measurement
文氏管，文丘里管

original *adj.* /əˈrɪdʒənl/   existing at the beginning of a particular period, process or activity
原来的

flatten *v.* /ˈflætn/   to become or make sth become flat or flatter
把……弄平

granted *conj.* /ˈgrɑːntɪd/   if, given that 假设

protrusion *n.* /prəˈtruːʒn/   a thing that sticks out from a place or surface
突出部分

layer *n.* /ˈleɪə(r)/   a quantity or thickness of sth that lies over a surface or between surfaces
层

imbalance *n.* /ɪmˈbæləns/   a situation in which two or more things are not the same size or are not treated the same
不平衡

leading edge   前缘
trailing edge   后缘
chord line   翼弦线
relative wind   相对风
angle of attack   迎角，攻角
relate to   有关
pressure gradient   压力梯度

Notes

1. Newton's First Law of Motion (Inertia) 牛顿第一定律

An object at rest remains at rest, and an object in motion remains in motion at constant speed and in a straight line unless acted on by an unbalanced force.

2. Newton's Second Law of Motion (Force) 牛顿第二定律

The acceleration of an object depends on the mass of the object and the amount of force applied.

3. Newton's Third Law of Motion (Action & Reaction) 牛顿第三定律

Whenever one object exerts a force on another object, the second object exerts an equal and opposite on the first.

4. Bernoulli's Principle 伯努利效应

In fluid dynamics, Bernoulli's principle states that an increase in the speed of a fluid occurs simultaneously with a decrease in static pressure or a decrease in the fluid's potential energy. The principle is named after Daniel Bernoulli who published it in his book *Hydrodynamica* in 1738.

Exercises

I. Read the passage and answer the following questions.

1. What are the four forces acting upon an airplane in flight?
2. What does airfoil refer to in this paragraph?
3. What is the angle of attack? What is its role?
4. How does a wing actually create lift?
5. What do you known about Newton's Three Laws of Motion and Bernoulli's Principle?

II. Multiple Choice (one or more answers).

1. When an aircraft is in straight and level flight, _____.
   A. an aircraft is maintaining its heading, altitude and airspeed
   B. lift equals weight
   C. thrust equals drag
   D. all of the above

2. Which of the following about fluid is true?
   A. It is a liquid.
   B. It deforms under an applied stress.
   C. It equals airfoil.
   D. It is a part of wing.

3. What do we know about the cord line?
   A. It is a real line on the wing.
   B. It can connect two edges of the wing.
   C. It forms an angle with the relative wind.
   D. The longer the line is, the more life is generated.

4. What is the major factor that affect the amount of lift?
   A. The angle of attack.
   B. The wings cord line.
   C. The leading edge.
   D. The trailing edge.

Text B Extensive Reading

Principles of Flight (2)

Now that we've covered the two theories behind lift, let's discuss all the factors that **determine** how much lift is produced. The best way to discuss this is through the lift **equation**. Don't worry though, this isn't a math lesson. Lift equals one-half **times** the **air density**, times the **surface area** of the wing, times the airplanes velocity **square**d, times the **coefficient** of lift. For the most part, this should be fairly straightforward. The only one that might confuse you is coefficient of lift. The coefficient of lift is simply just a number that is associated with a particular shape of an airfoil as well as the airfoils angle of attack.

Generally speaking, there are really only two ways a pilot can control the amount of lift the wings can generate: airspeed or angle of attack. The faster the airplane travels, the more lift the wings will generate. Similarly, the higher the airplanes angle of attack, the more lift the wings will generate. However, there's a limit to this angle. Let's look at this using a **chart**. The chart is **plot**ting the coefficient of lift of a particular wing. As its angle of attack increases, lift will continue to increase until a certain angle of attack, called the **critical angle of attack**. After this point, the wings will still create lift, but the amount of lift created is decreased. This is called a stall.

In addition to the aircraft's airspeed and angle of attack, there are other factors that affect the amount of lift created by the wings. However, the pilot does not have control over these factors. These factors have to do with the design of the wing itself and consist of the wing's **planform**, camber, aspect ratio and wing area. The wing's planform refers to the shape of the wing when viewed from above. The camber is the **curvature** of the wing. A wing with zero camber would be considered **symmetrical** about the **chord line**. **Camber** is usually designed into an airfoil to increase the maximum coefficient of lift, and thereby minimizing the **stall** speed of the aircraft. The wings **aspect ratio** is the relationship between the length and width of the wing. Generally, the higher the aspect ratio, the more efficient the generation of lift is. For example, **glider**s have really long **skinny** wings, giving them a higher aspect ratio compared to a Cessna 172. Finally, there is the

determine v. /dɪˈtɜːmɪn/ 决定

equation n. /ɪˈkweɪʒn/ 方程

times prep. /taɪmz/ 乘以

air density 空气密度

surface area 表面积

square adj. /skweə(r)/ 平方

coefficient n. /ˌkəʊɪˈfɪʃnt/ 系数

chart n. /tʃɑːt/ 图表

plot v. /plɒt/ 描绘；绘制

critical angle of attack 临界迎角

planform n. /ˈplænfɔːm/ 平面形状，俯仰图

curvature n. /ˈkɜːvətʃə(r)/ 弯曲度

symmetrical adj. /sɪˈmetrɪkl/ 对称的

chord line 翼弦线

camber n. /ˈkæmbə(r)/ 弧度，曲度

stall n. /stɔːl/ （飞机的）失速

aspect ratio 纵横比，展弦比

glider n. /ˈɡlaɪdə(r)/ 滑翔机

skinny v. /ˈskɪni/ 极窄的

wing area which is simply the total surface area of the wing's. The larger the wing area, the more lift the wing can produce.

Looking back at the lift equation, we can see that the wing area is **incorporate**d into the equation. The rest of the wing shape factors are **merge**d into the coefficient of lift **variable**. We saw how pilots can control the lift generated by the wings by changing the aircraft's airspeed and angle of attack.

However, most airplanes come equipped with one or more additional ways for the pilot to **manipulate** the wings and **in essence** change the shape of the wings. These are called **high-lift devices**, the most common of which are trailing edge flaps or just flaps for short. High-lift devices, such as flaps are designed to increase the lift and drag generated by the wings at low air speeds. Flaps are particularly important for the **approach** and **landing phase**s of flight. Use of flaps during a landing allows the pilot to fly at a fairly **steep descent angle** without gaining airspeed and allows the airplane to **touchdown** at a much slower airspeed. Flaps can generally be lowered in steps or more **precisely** in set degree amounts. **Initially** the input of flaps will increase lift by a larger amount with only a small increase in drag. As the flaps are **extend**ed further, usually around the **halfway point**, lift increases only slightly and the amount of drag created increases rapidly.

Now that we have an idea on how lift is generated, let's discuss the three remaining forces, starting with weight. Weight is the force of **gravity** pulling the aircraft back down to the earth. This force always acts **vertically downward** to the center of the earth no matter what the aircraft's attitude. The weight force always extends and **pivot**s from the center of gravity, also known as its CG. Keep in mind that the weight of an aircraft is not **constant**. It will vary with the equipment that is **install**ed as well as the passengers, cargo and fuel. Throughout the flight, the weight will slowly be decreasing as fuel is burned to **power** the engine.

Next is thrust. Thrust is the forward acting force opposing drag, which **propel**s the airplane through the air. In most general aviation airplanes, thrust is generated from the propeller. Larger **jet**s get their thrust from their turbine engines. Similar to lift, thrust is generated from the same principles as lift, but in a **horizontal** direction. A propeller is an airfoil as such as it rotates its **blades accelerate** the surrounding air towards the aft end of the aircraft. And as **illustrate**d with Newton's Third Law, the equal and opposite reaction results in the aircraft moving forward.And

wing area 机翼面积

incorporate v. /ɪnˈkɔːpəreɪt/ 包含

merge v. /mɜːdʒ/（使）合成

variable adj. /ˈveəriəbl/ 可变的

manipulate v. /məˈnɪpjʊleɪt/ 操纵

in essence 在本质上

high-lift device 高升力装置，增升装置

approach n. /əˈprəʊtʃ/ 进近

landing phase 着陆阶段

steep descent angle 急剧下降的角度

touchdown n. /ˈtʌtʃdaʊn/ 着陆

precisely adv. /prɪˈsaɪsli/ 精确地

initially adv. /ɪˈnɪʃəli/ 最初

extend v. /ɪkˈstend/ 展开，扩展

halfway point 半程处

gravity n. /ˈɡrævəti/ 地心引力

vertically downward 垂直向下的

pivot v. /ˈpɪvət/（使）旋转

constant adj. /ˈkɒnstənt/ 恒定的

install v. /ɪnˈstɔːl/ 安装

power v. /ˈpaʊə(r)/ 驱动

propel v. /prəˈpel/ 推动

jet n. /dʒet/ 喷气机

horizontal adj. /ˌhɒrɪˈzɒntl/ 水平的

blades n. /bleɪdz/ 叶片

finally, we reach our last force, drag. Drag is the force opposing thrust, which limits the forward speed of an aircraft. There are two types of drag: parasite and induced drag. **Parasite drag** is a direct result of the air resistance as the airplane flies through the air. There are three types of parasite drag: form drag, interference drag and skin friction drag.

accelerate *v.* /ək'seləreɪt/ 加速

illustrate *v.* /'ɪləstreɪt/ 说明

parasite drag 寄生阻力，废阻力

![Figure 3.3 Induced Drag diagram showing: A. High pressure air joins low pressure air at the trailing edge of the wing and wingtips. B. Wingtip vortices develop. C. The downwash increases behind the wing. D. The average relative wind is inclined downward and rearward, and lift is inclined aft. The rearward component of lift is induced drag.]

Figure 3.3 Induced Drag

The amount of parasite drag varies with the speed of the aircraft. As the airplane speed increases, the amount of parasite drag will increase. In fact, the amount of parasite drag (that) you experience is directly **proportional** to the square of the airspeed. For example, an aircraft traveling at 120 **knot**s will experience four times as much parasite drag as the same plane going 60 knots at the same altitude.

proportional *adj.* /prə'pɔːʃənl/ 成比例的

knot *n.* /nɒt/ 节（飞行器的速度计量单位）

The other kind of drag is **lift-induced drag**, more commonly called induced drag. While the wing is creating lift, behind the wing is a **downwash** of air. At the same time, the airflows around the wingtips are creating **vortices** that **spiral** from below the wing to above the wing. As these vortices wrap around the wing, they actually change the downwash angle of the air flowing over the wing. This in effect **tilt**s the direction of the lift created backwards. This **shift** from completely vertical lift to slightly aft lift is due to induced drag. Induced drag is higher at slower air speeds and decreases as we increase speed. This is because induced drag is worse when the airplane is flying at a high angle of attack, like when we are flying slowly.

lift-induced drag 诱导阻力

downwash *n.* /'daʊnwɒʃ/ 气流下洗

vortices *n.* /'vɔːtɪˌsiːz/ 涡流

spiral *v.* /'spaɪrəl/ 盘旋上升

tilt *v.* /tɪlt/（使）倾斜

shift *v.* /ʃɪft/ 移动

One way that a pilot can experience reduced induced drag is by flying the ground effect. When flying within a **wingspan** of the ground, the ground itself changes the downwash of the air flowing over the wings. This shifts the lift **vector** forward and reduces the amount of induced drag. Pilots can take advantage of the ground effect when performing a soft field takeoff. This lets the airplane lift off the ground before the regular liftoff speed. However, they'll need to **hover** over the ground for a few seconds to increase their speed before they can continue to climb out.

If we take both induced and parasite drag, plot them on a **graph** and add them together, we get a new **curve** representing total drag. The lowest point on the total drag curve shows the air speed at which we make the most amount of lift and the least amount of drag. This value is called L over D Max, or as pilots know it, our best glide speed. Pilots should be familiar with this number because in the unlikely event of an engine failure, this is the speed at which they'll want to glide down to the ground. And in a no wind condition, this speed will give the pilot the best glide ratio, meaning that they'll be able to stay **aloft** the longest to **maneuver** to their intended field for emergency landing.

Note on the left side of the total drag curve, the slower you fly, the more drag you create. In this region of air speeds, sometimes called the backside of the power curve, the pilot will actually have to add more and more thrust to counter the high amounts of drag being created. In fact, if they want to accelerate out of this range of air speed, they will have to add an excessive amount of power, maybe even full power.

One other thing to keep in mind at slow air speeds is that there is much less airflow traveling over the flight control surfaces. As such, any input you make on the flight controls will not have the fast response one would be used to. The flight controls will feel **mushy** and they may require large inputs before any real response is felt.

While the finer details of the principles of flight can seem a bit **overwhelming** at first gaining, a basic knowledge of how the airplane flies provides the pilot full understanding of all the forces at work and the best methods and techniques for controlling their aircraft. A good pilot doesn't just drive their plane from point A to B, but instead understands the art and science of how their plane flies.

wingspan *n.* /ˈwɪŋspæn/ 翼展

vector *n.* /ˈvektə(r)/ 矢量

hover *v.* /ˈhɒvə(r)/ 盘旋

graph *n.* /ˈgraːf/ 图
curve *n.* /ˈkɜːv/ 曲线

aloft *adv.* /əˈlɒft/ 在高空
maneuver *v.* /məˈnʊvə/ 机动

mushy *adj.* /ˈmʌʃi/ 糊里糊涂的

overwhelming *adj.* /ˌəʊvəˈwelmɪŋ/ 巨大的

Exercises

I. Answer the questions.

1. What is the critical angle of attack?
2. What are the factors that affect the amount of lift created by the wings and can not be controlled by the pilot, and why?
3. What are flaps used for? how?
4. Where is thrust generated?
5. What do you know about the parasite drag, and the induced drag?
6. What is L over D Max?

II. Multiple choice (one or more answers).

1. Which of the following can help you know the amount of lift generated?
   A. Air density.
   B. The surface area of the wing.
   C. The airplane's velocity.
   D. All of the above.
2. Why can't the angle of attack increase indefinitely?
   A. Because after critical angle of attack, the wings will still create lift, but the amount of lift created is decreased.
   B. Because after critical angle of attack, the friction drag will increase.
   C. Because after critical angle of attack, the top is rounded while the bottom is relatively flat.
   D. Because after critical angle of attack, it rotates its blades accelerate the surrounding air.
3. Which of the following statements about weight is true?
   A. Weight is the force of gravity pulling the aircraft back down to the earth.
   B. This force always acts vertically upward to the center of the earth.
   C. The weight of an aircraft is the same from departure to landing.
   D. The weight of an aircraft increases as fuel is burned.
4. Which of the following statement about the factors that affect the amount of lift is true?
   A. The wing's planform refers to the shape of the wing when viewed from below.
   B. Camber should be designed to be symmetrical about the chord line.
   C. The higher the aspect ratio, the more efficient the generation of lift is.
   D. The wing area has nothing to do with the lift produced.
5. Why is induced drag reduced by flying into the ground?
   A. Because the angle of attack decreases, the induced drag decreases.
   B. Because lift increases only slightly and the amount of drag created decreases rapidly.
   C. Because the airflows around the wingtips are creating vortices that spiral from below to above the wing.
   D. Because the ground changes the downwash of the air flowing over the wings.

Part 3 Translation

1. Lift is a force that is produced by the dynamic effect of the air acting on the airfoil, and acts perpendicular to the flight path through the center of lift and perpendicular to the lateral axis.

2. Weight is a force that pulls the aircraft downward because of the force of gravity. It opposes lift and acts vertically downward through the aircraft's center of gravity.

3. In normal flight, as air flows around the wing, the air gets deflected downward as it flows smoothly around it, and as a result the wind will lift the wing up.

4. Wings are designed to create this kind of pressure gradient, because air always moves from areas of high pressure to areas of low pressure. Since the wing is stuck in between the two areas of unequal pressure, it is lifted towards the area of low pressure by the force of the higher pressure air trying to move to the low pressure side of the wing.

5. Generally speaking, there are really only two ways a pilot can control the amount of lift the wings can generate: airspeed or angle of attack. The faster the airplane travels, the more lift the wings will generate. Similarly, the higher the airplane's angle of attack, the more lift the wings will generate.

6. 像襟翼这样的增升装置（高升力装置）是为了在低空速条件下增加机翼产生的升力和阻力。

7. 在整个飞行过程中，由于燃料的消耗，重力将慢慢下降。

8. 推力产生的原理与升力相同，但推力是水平作用力。

9. 如果我们将诱导阻力和寄生阻力画在同一张图上，并把它们加在一起，我们就得到了一条代表总阻力的新曲线。

10. 虽然刚开始了解飞行原理的细节可能会让人不知所措，但了解飞机如何飞行的基本知识能让飞行员充分理解所有起作用的力，以及控制飞机的最佳方法和技术。

Part 4 Supplementary Reading

Ground Effect

Ever since the beginning of manned flight, pilots realized that just before touchdown it would suddenly feel like the aircraft did not want to go lower, and it would just want to go on and on. This is due to the air that is trapped between the wing and the landing surface, as if there were an air cushion. This phenomenon is called ground effect.

When an aircraft in flight comes within several feet of the surface, ground or water, a change occurs in the three dimensional flow pattern around the aircraft because the vertical component of the airflow around the wing is restricted by the surface. This alters the wing's upwash, downwash, and wingtip vortices(Figure 3.4). Ground effect, then, is due to the interference of the ground (or water) surface with the airflow patterns about the aircraft in flight. While the aerodynamic characteristics of the tail surfaces and the fuselage are altered by ground effect, the principal effects due to proximity of the ground are the changes in the aerodynamic characteristics of the wing. As the wing encounters ground effect and is maintained at a constant AOA, there is consequent reduction in the upwash, downwash, and wingtip vortices.

Figure 3.4 Ground Effect Changes Airflow

Induced drag is a result of the airfoil's work of sustaining the aircraft, and a wing or rotor lifts the aircraft simply by accelerating a mass of air downward. It is true that reduced pressure on top of an airfoil is essential to lift, but that is only one of the things contributing to the overall effect of pushing an air mass downward. The more downwash there is, the harder the wing pushes the mass of air down. At high angles of attack, the amount of induced drag is high; since this corresponds to lower airspeeds in actual flight, it can be said that induced drag predominates at low speed. However, the reduction of the wingtip vortices due to ground effect alters the spanwise lift distribution and reduces the induced AOA and induced drag. Therefore, the wing will require a lower AOA in ground effect to produce the same CL. If a constant AOA is maintained, an increase in CL results (Figure 3.5).

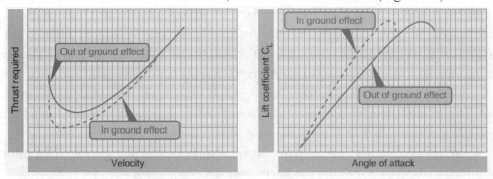

Figure 3.5 Ground Effect Changes Drag and Lift

Ground effect also alters the thrust required versus velocity. Since induced drag predominates at low speeds, the reduction of induced drag due to ground effect will cause a significant reduction of thrust required (parasite plus induced drag) at low speeds. Due to the change in upwash, downwash, and wingtip vortices, there may be a change in position (installation) error of the airspeed system associated with ground effect. In the majority of cases, ground effect causes an increase in the local pressure at the static source and produces a lower indication of airspeed and altitude. Thus, an aircraft may be airborne at an indicated airspeed less than that normally required.

In order for ground effect to be of significant magnitude, the wing must be quite close to the ground. One of the direct results of ground effect is the variation of induced drag with wing height above the ground at a constant CL. When the wing is at a height equal to its span, the reduction in induced drag is only 1.4 percent. However, when the wing is at a height equal to one-fourth its span, the reduction in induced drag is 23.5 percent and, when the wing is at a height equal to one-tenth its span, the reduction in induced drag is 47.6 percent. Thus, a large reduction in induced drag takes place only when the wing is very close to the ground. Because of this variation, ground effect is most usually recognized during the liftoff for takeoff or just prior to touchdown when landing.

During the takeoff phase of flight, ground effect produces some important relationships. An aircraft leaving ground effect after takeoff encounters just the reverse of an aircraft entering ground effect during landing. The aircraft leaving ground effect will:
· Require an increase in AOA to maintain the same CL
· Experience an increase in induced drag and thrust required
· Experience a decrease in stability and a nose-up change in moment

· Experience a reduction in static source pressure and increase in indicated airspeed

Ground effect must be considered during takeoffs and landings. For example, if a pilot fails to understand the relationship between the aircraft and ground effect during takeoff, a hazardous situation is possible because the recommended takeoff speed may not be achieved. Due to the reduced drag in ground effect, the aircraft may seem capable of takeoff well below the recommended speed. As the aircraft rises out of ground effect with a deficiency of speed, the greater induced drag may result in marginal initial climb performance. In extreme conditions, such as high gross weight, high density altitude, and high temperature, a deficiency of airspeed during takeoff may permit the aircraft to become airborne but be incapable of sustaining flight out of ground effect. In this case, the aircraft may become airborne initially with a deficiency of speed and then settle back to the runway.

A pilot should not attempt to force an aircraft to become airborne with a deficiency of speed. The manufacturer's recommended takeoff speed is necessary to provide adequate initial climb performance. It is also important that a definite climb be established before a pilot retracts the landing gear or flaps. Never retract the landing gear or flaps prior to establishing a positive rate of climb and only after achieving a safe altitude.

If, during the landing phase of flight, the aircraft is brought into ground effect with a constant AOA, the aircraft experiences an increase in CL and a reduction in the thrust required, and a "floating" effect may occur. Because of the reduced drag and lack of power-off deceleration in ground effect, any excess speed at the point of flare may incur a considerable "float" distance. As the aircraft nears the point of touchdown, ground effect is most realized at altitudes less than the wingspan. During the final phases of the approach as the aircraft nears the ground, a reduction of power is necessary to offset the increase in lift caused from ground effect otherwise the aircraft will have a tendency to climb above the desired glidepath (GP).

Unit 4 Aircraft Performance and Limitations

Lead-in

This chapter discusses the factors that affect aircraft performance while taking off, climbing, cruising, descending and landing, which include the aircraft weight, atmospheric conditions, and runway environment acting on an aircraft.

This unit will cover:
· How weight of the airplane affects aircraft performance
· How density of the air affects aircraft performance
· How environment factors, such as humidity, temperature, altitude pressure, and wind affect aircraft performance

Part 1 Listening and Speaking

New Words & Expressions

freighter *n.* /ˈfreɪtər/   plane that carries goods
货机，运输机

reject *v.* /rɪˈdʒekt/   stop   中断
decline *v.* /dɪˈklaɪn/   refuse   拒绝
inspection *n.* /ɪnˈspekʃn/   check on   检查
initiate *v.* /ɪˈnɪʃieɪt/   start doing or enforcing something
开始实施

sustain *v.* /səˈsteɪn/   suffer from   遭受
occurrence *n.* /əˈkɜːrəns/   incidents, things that happen   事件
tail strike phrasal   尾部擦地
balked landing   中止的着陆，不成功的着陆

I. Listen to the passage and fill in the blanks with the words you hear.

On August 11, 2021, A UPS United Parcel Service MD-11 freighter, performing flight 5X-2876 from Louisville to Phoenix (USA), landed on Phoenix Sky Harbor's runway 25L but _____ landing and went around, tower immediately queried whether any assistance was needed, the crew _____. The crew of another aircraft _____ to tower the MD-11 just had a _____ strike. A runway _____ was _____, nothing was found on the runway and the runway was returned to service. The MD-11 positioned for another approach to runway 25R and landed without further incident about 10 minutes after the _____ landing. The FAA reported:"AIRCRAFT _____ A TAIL STRIKE DURING LANDING AT PHOENIX." The damage was minor. The _____ was _____ an incident.

II. Filling the blanks with the words from the box.

① density	② performance	③ rates
④ calculate	⑤ critical	⑥ consumption

Aircraft manufacturers publish _____ charts to allow pilots to _____ the aircraft's performance numbers, including things such as takeoff and landing distances, climb _____, true airspeed and fuel _____. The two major factors that affect an aircraft's performance are the weight of the airplane and the _____ of the air the airplane is flying through. Additionally, the wind can play a factor, especially in the most _____ phases of flight: takeoff and landing.

III. Read the dialogue and Discuss.

C – Air Traffic Controller   P – Pilots
P: B Tower, HDA305, ready for takeoff.

C: HDA, cleared for takeoff runway 12.

P: Cleared for takeoff, HDA.

P: Tower, HDA, we have suffered a tail strike on takeoff.We intend to return for landing.

C: HDA, do you know the reason for tail strike?

P: Affirm. It results in low takeoff speeds. And I regret that we did not have cross-checking of the speeds during the departure.

C: Do you think your aircraft has any damage?

P: I am not sure. Perhaps the aft pressure bulkhead has been severely damaged.

Discussion

1. At which phases of flight is tail strike most likely to occur?
2. What are the possible reasons for tail strike?
3. What will you do after tail strike?
4. What can be done to avoid tail strike?

Part 2 Reading

Text A Intensive Reading

Aircraft Performance and Limitations (1)

There are thousands of different types of aircraft flying around in the skies today, from **commercial** airliners that carry hundreds of people to and from their **destination**s to small single-seat **aerobatic** airplanes that perform amazing **feat**s of **aerial agility**. Regardless of what role the manufacturer designed them for, all planes have one thing in common – how well they perform is directly affected by the environment in which they fly. In this segment we will look at various factors that determine aircraft performance and how pilots can plan in advance for **variation**s in that performance by using the information provided by the **manufacturer** for their specific airplane.

Figure 4.1 Factors Affecting Performance

Aircraft manufacturers publish performance charts to allow pilots to **calculate** the aircraft's performance numbers, including things such as takeoff and landing distances, climb rates, true airspeed and **fuel consumption**. The two major factors that affect an aircraft's performance are the weight of the airplane and the **density** of the air the airplane is flying through. Additionally, the wind can play a factor, especially in the most **critical phase**s of flight: takeoff and landing.

We will first look at how weight affects aircraft performance. If we consider an airplane at level flight, the wings are producing the same amount of lift as the weight of the airplane. If the plane weighs more, the amount of lift the wings have to produce increases as well. A **byproduct** of lift is **drag**, so if the amount of lift is increased, the amount of drag will also increase. The increased drag will result in a slower airspeed for a given power **setting**, thus decreasing the **cruise** performance of the airplane. So, if we **load** our airplane to its maximum **gross weight**, we will get a slower cruise speed than if we were flying at a lighter weight.

Just as important as the weight of the airplane is the density of the air. Air density is a measure of how far air **molecule**s are **space**d apart. The closer the molecules are spaced, the denser the air is. The density of the air is mainly affected by three factors, those being pressure, temperature and **humidity**. Pressure affects how many air particles are **squeeze**d into a given volume of air. The less pressure applied to a parcel air means the fewer air molecules it will contain, thereby resulting in a decrease in air density. Temperature also has a direct effect on air density. The higher the temperature of the air, the more space the molecules take up as they move around. This means that higher temperatures result in a decrease in air density.

One last factor to consider is humidity. Humidity alone is not considered as an **essential** factor in calculating density altitude and aircraft performance; however, it does contribute. The higher the humidity of the air, the less dense the air is. This is because the water vapor is lighter than air. So, the more water vapor in the air, the lower the resulting density will be. Expect a decrease in overall performance in high humidity conditions. Also, the pressure of the air goes down as altitude increases, so the higher you are flying, the lower the air pressure will be, and consequently the lower the density will be.

A better way for pilots to **make sense of** how density affects airplane performance is to use something called density altitude. Density altitude is a way of describing how the airplane will perform when put in context of an altitude. Scientists have developed a **formula** that allows them to **assign** a specific value for temperature and pressure to different altitudes. We call these designated values the standard **atmospheric** conditions for those altitudes. The standard conditions for 0 feet **elevation** or sea level consist of a temperature of 15 degrees **Celsius** and a pressure of 29.92 inches of **mercury.** Let's take an example of a Cessna departing from an airport at sea level. If the temperature is 15 degrees Celsius and the pressure is 29.92 inches of mercury, the plane would perform as if it's flying at a density altitude of 0 feet. If that same airplane were to depart from the same airport on a different day with different weather conditions, we would use the NORM standard temperature and pressure to determine what density altitude those conditions would represent. For instance, if the temperature was 32 degrees Celsius and the pressure was 30 inches of mercury, this would be the **equivalent** of a density altitude of approximately 2,100 feet. The pilot would then know to expect the same performance he or she would get at 2,100 feet instead of sea level. This would result in the plane using up more runway to take off and having a slower climb up to the desired cruising altitude.

Performance charts from aircraft manufacturers are published, so pilots can take the changing weather conditions into account when they're calculating aircraft performance. These charts provide a simple **graphical** way of determining density altitude without the pilot needing to perform complex math problems to get the same results.

Let's say we will be flying at 5,000 feet to our destination and want to find out what our density altitude will be. For simplicity, we'll use a density altitude chart. The first step involved is to calculate our pressure altitude, which corrects for our non-standard pressure. Let's say today's pressure is 30.1 inches of mercury. Looking at the right side of the chart, we look into the left column for the altimeter setting of three zero point one zero (30.10), and find that the pressure altitude **conversion factor** in the right column is negative 165. That means that our pressure altitude is 165 feet lower than what's indicated. 5,000 feet minus 165 feet equals a pressure altitude of 4,835 feet. If we were to enter two nine point nine two (29.92)in our altimeter Kollsman window, it would read 4,835 feet.

Now that we have our pressure altitude, we need to correct for non-standard temperature, giving us our density altitude. Let's say that the current outside temperature read of the air is about 4 degrees Celsius, starting at the bottom of the chart find the 4-degree mark. Next follow the **vertical** lines upward until we **intersect** the **diagonal** pressure altitude of 4,835 feet. Then, follow the horizontal lines to the left edge and read the altitude. That would result in the approximate altitude of 4,600 feet. Today's relatively high pressure and low temperature resulted in a lower density altitude, which means the plane will perform as if it were flying at a lower altitude, giving it an increase in performance.

So now that we are able to determine pressure and density altitudes, we can take a closer look at how this affects the airplane.The Pilots Information Manual contains performance charts for the different phases of flight, including takeoff, climb, cruise and landing. Remember that each aircraft manufacturer creates their performance charts differently, so be sure to study them carefully before you actually use them.

Let's start off by looking at the short field takeoff distances for 2,400 pounds. The distances shown are for **ground roll** and for clearing a 50-foot **obstacle**. The aircraft's ground roll is how far down the runway the airplane must travel to **accelerate** from a standing start to a **rotation speed** of 51 knots. The column labeled total feet to clear a 50-foot obstacle gives the distance the plane will travel, accelerating from a standing start to take off and then climb to an altitude of 50 feet. On the chart, the far left column shows the pressure altitude, and the top row shows temperature. In essence, the chart will calculate the effects of density altitude for you. Let's say, for example, our pressure altitude as sea level and the temperature is 0 degrees Celsius. If we look at the intersection of these two values, we find that the take-off distance is 745 feet. As the temperature increases, notice how the take-off distance increases. At 20 degrees Celsius, the take-off distance increases to 860 feet; and at 40 degrees Celsius, the take-off distance increases to 995 feet. This results in a total difference of two hundred and fifty feet increased just from the temperature rising.

Now let's look at altitude changes. A sea-level pressure altitude and a zero degree temperature result(s) in a take-off distance of 745 feet. At a pressure altitude of 5,000, like you would find ERAU's Prescott Arizona campus, the takeoff roll is 1,170 feet. That results in a difference of 425 feet. Now imagine a hot summer day with a pressure altitude of 5,000 feet and a temperature of 30 degrees Celsius, the take-off distance in that situation would be increased to 1,465 feet. That's nearly double the distance compared to what it would have been at sea level and zero degrees.

performance *n.* /ˌpəˈfɔːməns/   how successful someone is or how well he does something.
业绩，性能

limitation *n.* /ˌlɪmɪˈteɪʃn/   the act or process of controlling or reducing it.
限制

commercial *adj.* /kəˈmɜːrʃl/   involving or relating to the buying and selling of goods
商业的

destination *n.* /ˌdestɪˈneɪʃn/   the place to which they are going or being sent
目的地

aerobatic *adj.* /ˌerəˈbætɪk/   spectacular flying feats and maneuvers (such as rolls and dives)
[航] 用特技飞行的

feat *n.* /fiːt/   a deed notable especially for courage
壮举，技艺表演

aerial *adj.* /ˈeriəl/   of, relating to, or occurring in the air or atmosphere;of or relating to aircraft
空中的，从飞机上的

agility *n.* /əˈdʒɪləti/   the quality or state of being agile
敏捷性

variation *n.* /ˌveriˈeɪʃn/   a change or slight difference in a level, amount, or quantity
变化，差别

manufacturer *n.* /ˌmænjʊˈfæktʃərə(r)/

a business or company which makes goods in large quantities to sell

[商业] 生产商

calculate *v.* /ˈkælkjʊleɪt/

discover it from information that you already have, by using arithmetic, mathematics, or a special machine

计算

fuel *n.* /ˈfjuːəl/

a substance such as coal, oil, or petrol that is burned to provide heat or power

燃料

consumption *n.* /kənˈsʌmpʃn/

the act of using them or the amount used

消耗

density *n.* /ˈdensəti/

the extent to which something is filled or covered with people or things

密度

critical *adj.* /ˈkrɪtɪkl/

relating to or denoting a point of transition from one state to another

临界的

phase *n.* /feɪz/

a particular stage in a process or in the gradual development of something.

阶段

byproduct *n.* /ˈbaɪˌprɑːdʌkt/

something that is produced during the manufacture or processing of another product

副产品

drag *v.* /dræg/

the retarding force acting on a body (such as an airplane) moving through a fluid (such as air) parallel and opposite to the direction of motion

阻力

setting *n.* /ˈsetɪŋ/

the manner, position, or direction in which something is set

设置

cruise *n.* /kruːz/

travel around it or along it on a cruise.

巡航

load *v.* /ləʊd/

put a large quantity of things into a vehicle or a container

大量装入

molecule *n.* /ˈmɑːlɪkjuːl/   the smallest amount of a chemical substance which can exist by itself
分子

space *v.* /speɪs/   postion at a distance from one another
把……分隔开

spacing *n.* /ˈspeɪsɪŋ/   space between things
间隔

humidity *n.* /hjuːˈmɪdəti/   the amount of water in the air
湿度

squeeze *v.* /skwiːz/   get the liquid or substance out by pressing the object
挤，压榨

essential *adj.* /ɪˈsenʃl/   the most basic or important aspects of something
基本的，重要的

formula *n.* /ˈfɔːrmjələ/   a group of letters, numbers, or other symbols which represents a scientific or mathematical rule
公式

assign *v.* /əˈsaɪn/   to appoint as a duty or task
分配

atmospheric *adj.* /ˌætməsˈfɪrɪk, ˌætməsˈferɪk/
relating to the Earth's atmosphere
大气层的

elevation *n.* /ˌelɪˈveɪʃn/   height above a given level, esp. sea level
高度，海拔

celsius *n.* /ˈselsiəs/   relating to, conforming to, or having the international thermometric scale on which the interval between the triple point of water and the boiling point of water is divided into 99.99 degrees with 0.01° representing the triple point and 100° the boiling point
摄氏温度

mercury *n.* /ˈmɜːrkjəri/   the column of mercury in a thermometer or barometer
（温度计的）水银柱

equivalent *n.* /ɪˈkwɪvələnt/   a thing, amount, word, etc. that is equivalent to sth else

	相等的东西，等量，对应词
graphical *adj.* /ˈɡræfɪkl/	of something uses graphs or similar images to represent statistics or figures 图解的
conversion *n.* /kənˈvɜːrʒn/	the act or process of changing something into a different state or form (状态或形式的) 改变
factor *n.* /ˈfæktər/	a particular level on a scale of measurement 系数
vertical *adj.* /ˈvɜːrtɪkl/	going straight up or down from a level surface or from top to bottom in a picture, etc. 垂直的
intersect *v.* /ˌɪntərˈsekt/	meet or cross each other 相交
diagonal *n.* /daɪˈæɡənl/	a diagonal line or movement going in a sloping direction, for example, from one corner of a square across to the opposite corner 对角线
obstacle *n.* /ˈɑːbstəkl/	an object that makes it difficult for you to go where you want to go, because it is in your way 障碍物
accelerate *v.* /əkˈseləreɪt/	to start to go faster 加速
gross weight	毛重
make sense of	弄清……的意思
ground roll	着陆或起飞滑跑
rotation speed	旋转速度，转速

Notes

1. **Kollsman window** 高度表气压调定窗

Barometric Reading. The Kollsman window is located at the 3 o'clock position on the altimeter dial. This window allows access to read a sub-dial, which contains the barometric readings. The arrowhead indice located precisely at the 3 o'clock position on the altimeter's main dial is used as the reference point for reading the barometric sub-dial. Most altimeters will have a sub-dial, which covers the readings from 28.1 InHg (inches of mercury) to 31.0 InHg. On the sub-dial each major indice is read as 0.1 InHg, each minor indice is read as 0.02 InHg.Typically pilots will obtain a local barometric reading from the nearest airport. They will then set the Kollsman window to the setting that they received. This action will adjust the altimeter reading, eliminating error due to local weather conditions.

2. ERAU's Prescott Arizona 亚利桑那州普雷斯科特的安柏瑞德航空大学

Embry-Riddle Aeronautical University, Prescott is a residential campus of Embry-Riddle Aeronautical University in Prescott, Arizona. The university offers bachelor, master's, and PhD degree programs in arts, sciences, aviation, business, engineering, and security & intelligence.

Exercises

I. Answer the following questions.

1. What are the performance charts published by manufacturers for?
2. How does weight affect aircraft performance?
3. How does air density affect aircraft performance?
4. What are the factors that affect the air density? How do they affect aircraft performance respectively?

II. Multiple choice (one or more answers).

1. What is this passage mainly talking about?
   A. The factors affecting aircraft performance and how they can be used in pre-flight planning.
   B. The environment of flight and pilots.
   C. Agility of aircraft and human factors.
   D. Commercial value of aircraft and aircraft manufacturers.
2. What are the factors that affect the performance of the plane?
   A. Humidity of environment.
   B. Weather.
   C. The weight of the aircraft.
   D. Density of air.
3. Which statement about density altitude is correct?
   A. Density altitude has no effect on aircraft performance.
   B. It's hard to calculate density altitude using the performance charts.
   C. A specific value for temperature and pressure is assigned to different altitudes.
   D. Pilots need to perform complex math problems to get the density altitude.
4. Which temperature corresponds to the wrong take-off distance?
   A. 0 degrees Celsius : 745 feet.
   B. 20 degrees Celsius : 860 feet.
   C. 40 degrees Celsius : 995 feet.
   D. 60 degrees Celsius: 1215 feet.

Text B Extensive Reading

Aircraft Performance and Limitations (2)

The aircraft's airspeed indicator reads off what we call indicated airspeed. This airspeed only takes the pressure of the air into account, instead of all of the density factors. Because of that, our true airspeed may be different from the indicated, depending on the temperature and humidity. On days when the air density is low, the aircraft will have to travel faster to get the same indicated airspeed. This results in increased takeoff distances. The wind direction and speed play a big role in takeoff distances as well. For the airplane to take off, it needs to reach a speed of 51 knots. If there is an 18-knot **head wind**, it's almost like saying the airplane has an 18-knot head start, and only needs to accelerate 32 more knots to get 51 knots of wind flowing over the wings. This would reduce the runway required from 745 feet down to only 596 feet. A tail wind would do just the opposite. If we were to have a mere of four-knot **tail wind**, the airplane would first have to travel at four knots just to **counteract** the wind. It would then have to continue accelerating an additional 51 knots to be able to take off. That means that over the ground the plane would have to accelerate up to 55 knots to take off. And while a four-knot increase doesn't sound like much, that little bit of wind would increase the **takeoff roll** to 894 feet, an increase of 150 feet.

head wind 逆风

tail wind 顺风

counteract v. /ˌkaʊntərˈækt/ 对······起反作用，抵消

takeoff roll 起飞滑跑

Another factor to consider in the performance of an airplane is its weight. Heavier airplanes require more power and thrust. Simply put the heavier the airplane is, the lower its performance would be. For takeoff distances, a heavier airplane would need more distance in their roll. This is because it takes more power to get the airplane rolling and accelerate to rotation speed. If we stick to our sea-level pressure altitude and 0 degree temperature, the 2,400 pound Cessna would require a take-off distance of 745 feet. If the weight was increased to 2,550 pounds, the take-off distance would increase to 860 feet. If we instead decreased our weight to 2,200 pounds, the distance would decrease to only 610 feet. For those 350 pounds, we'd get a difference in takeoff distance of 250 feet.

After our takeoff is accomplished, the next stage of flight is the climb. Our next chart not only shows the rate of climb that the airplane will get, but also includes the time, fuel and distance the airplane will use to get to our cruising altitude. Looking at this chart, we can understand the trends as an airplane climbs higher and higher. Notice in particular how the climb rates decrease with altitude. As the airplane climbs, the density decreases, and this decreases performance. So, at sea level, the best rate of climb speed is 74 knots, and the vertical climb rate is 730 feet per minute (FPM). At 8,000 feet, the best rate speed lowers to 72, and the climb rate drops to 410 feet per minute (FPM).

Temperature will also affect these numbers. The note at the bottom of the page states that all-time fuel and distance numbers increase by 10% for every 10 degrees above the standard temperature. Weight changes will also affect these performance values. While Cessna does not **publish** the performance values at lighter weights, if the airplane did weigh less than 2,550 pounds, the airplane should climb at a greater rate than the published numbers. However, since Cessna does not publish the information, we plan for the most conservative scenario at the 2,550-pound weight.

publish *v.* /ˈpʌblɪʃ/ 发表，公开

This next chart shows the cruise performance of the airplane, displaying the airspeed, fuel burn rate and percent of **maximum continuous power**, **in relation to** different temperatures and altitudes. At 4,000 feet with an engine RPM (Revolutions Per Minute) setting of 2500 RPM, and its standard temperature, in this case 7 degrees Celsius, the airplane will be at 69% power at 115 knots and use up 9.5 gallons per hour. If the temperature were to rise to 27 degrees, the power output would drop to 64%, the speed would reduce to 114 knots, and fuel burn would decrease to 8.9 gallons per hour. This happens without ever touching the **throttle**. The reduced air density gives both a drop in performance and a drop in fuel burn, because the engine has less air molecules available to use. If the temperature would have dropped to -13 degrees Celsius, the increased air density would **boost** the engine power to 74%, increase the speed to 115 knots, and consequently also increase the fuel burn to 10.1 gallons per hour. The different **power setting**s available at each altitude give the pilot more **flexibility** in flight planning, and choosing the one that best fits the flight plan.

maximum continuous power 最大连续功率
in relation to 关于，涉及

throttle *n.* /ˈθrɒtl/ 节流阀，油门杆，油门踏板

boost *v.* /buːst/ 促进
gallon *n.* /ˈgælən/ 加仑（液体单位，合 3.785 升）
power setting 功率设定
flexibility *n.* /ˌfleksəˈbɪləti/ 灵活性，航程和续航图

If we also look at the range and endurance charts, we can see how changing the RPM of the engine can greatly affect both how far we can travel and how long we can stay in the air. By simply going slower, we can actually travel over 100 miles further on one **tank** of gas. If your destination is about as far away as your current maximum range, it may be wise to slow down to a lower power setting. You'd be able to make it all the way there without needing to stop for fuel.

tank *n.* /tæŋk/（盛放液体或气体的）罐，箱，槽

Just like the takeoff, landing distances will also vary with the weight of the airplane, the outside air temperature and the airport **field elevation**. Cessna only publishes landing data for an airplane at the maximum gross weight of 2,550 pounds, but lighter airplanes we'll be able to fly a slower approach speed for landing, and we'll have less **momentum** and less energy to **dissipate** once on the runway, resulting in a shorter **rollout**. Just like before, make sure you use the published data to calculate your runway distance required before flight.

field elevation 机场标高

momentum *n.* /məʊˈmentəm/ 动量
dissipate *v.* /ˈdɪsɪpeɪt/ 驱散，消散
rollout *n.* /ˈrəʊlaʊt/ ［航］滑跑（飞机着陆时在跑道上减速的阶段）

Here is how temperature and pressure affect the landing distance in the Cessna 172. At sea level and at 0 degree Celsius, the Cessna will need 545 feet to land. As the temperature increases, the distances increase all the way to 625 feet at the 40-degree mark. Just like the takeoff performance charts, this chart shows the distance required not only for ground roll but also for the distance to clear a 50-foot obstacle, and then land, and come to a complete stop on the runway. The chart also shows the landing distances increase as altitude increases. Here, the distance of 545 feet at sea level increases to 655 feet when at 5,000 feet.

Wind will also affect the landing roll. A headwind of 9 knots will reduce the landing distance by 10%. And a tailwind of just 2 knots will increase the landing distance by 10%. So, landing at sea level with a temperature of zero and a headwind of 18 knots will reduce the landing distance from 545 feet down to 436 feet.

The examples we just used were based on the performance of a Cessna 172S, but the general concepts apply to all airplanes. Regardless of what you fly, weather conditions have a significant impact on aircraft performance. Doing your performance calculations prior to each flight is a critical part of your pre-flight planning. By understanding how the temperature, pressure and altitude can affect your aircraft's performance, you can ensure a safe and enjoyable flight every time you take to the air.

Exercises

I. Answer the following questions.

1. How is the indicated airspeed different from the true airspeed? What is its impact on the airplane's takeoff?
2. How does weight affect the aircraft performance during take-off?
3. As the airplane climbs, does the performance of aircraft decrease? Why?
4. What does the difference in the cruise performance of the airplane in relation to different temperatures and altitudes mean for the pilot?
5. If your destination is about as far away as your current maximum range, how should the RPM be set?
6. How do a headwind and a tailwind affect the landing roll?

II. Multiple choice (one or more answers).

1. What is wrong about the cruising performance of the aircraft?
   A. At 7 degrees Celsius,the airplane will be at 69% power at 115 knots and use up 9.5 gallons per hour.
   B. The reduced air density gives both a drop in performance and a drop in fuel burn.
   C. The different power settings available at each altitude give the pilot more flexibility in flight planning.
   D. At 27 degrees, the power output would drop to 80%, the speed would reduce to 114 knots, and fuel burn would decrease to 8.9 gallons per hour.
2. Which of the following affect the landing distance?
   A. The weight of the airplane.
   B. The airport field elevation.
   C. The wind.
   D. The outside air temperature.
3. Which statement is wrong about the landing distance?
   A. A tailwind will reduce the landing distance.
   B. A headwind will increase the landing distance.
   C. At sea level and at 0 degree Celsius, the Cessna will need 545 feet to land.
   D. The landing distance increases with the decrease of altitude.
4. What does pre-flight performance calculation NOT include?
   A. The temperature.
   B. The pressure.
   C. The composite.
   D. The humidity of environment.

Part 3 Translation

1. The two major factors that affect an aircraft's performance are the weight of the airplane and the density of the air the airplane is flying through. Additionally, the wind can play a factor, especially in the most critical phases of flight: takeoff and landing.

---

2. One factor to consider in the performance of an airplane is its weight. Heavier airplanes require more power and thrust. Simply put the heavier the airplane is, the lower its performance would be.

---

3. Air density is a measure of how far air molecules are spaced apart. The closer the molecules are spaced, the denser the air is.

---

4. 密度高度是一种在高度背景下描述飞机性能的方式。

---

5. 性能图表提供了一种简单的图形化方法来确定密度高度，而无需飞行员进行复杂的数学运算就能得出相同的结果。

---

6. 无论你驾驶的是什么飞机，天气状况都会对飞机的性能产生重大影响。每次飞行前进行性能计算是预飞行计划的关键部分。通过了解温度、压力和海拔对飞机性能的影响，你可以确保每次飞行都安全、愉快。

Part 4 Supplementary Reading

Underestimating the Importance of Weight and Balance

Many pilots, from sport pilot to commercial pilot, tend to underestimate the importance of proper weight and balance of their aircraft. Load sheets are taken for granted and hasty calculations are made of the aircraft's CG. Unfortunately, each year there are a number of accidents related to weight and balance issues. Many of these occurrences could have been avoided had more attention been given to weight and balance.

Every student pilot is taught how to work a weight and balance problem and that it is important to make sure every flight is loaded "within the envelope" (no more than maximum gross weight) for both takeoff and landing. But does he or she really understand just why this is so and the disastrous effect of being out of the envelope? Two examples of documented cases are provided below in an effort to indicate the serious nature of maintaining the proper weight and balance. In case studies when weight and balance was listed as the major factor of the accident, many were fatal.

For instance, a small aircraft was loaded with hunters, gear, and dogs (none of the dogs were secured inside the aircraft). During takeoff, all the dogs went to the aft of the airplane. This shifted the CG well aft of its allowable limit. The airplane stalled and crashed. The airplane was destroyed with casualties. Another accident occurred when a group of skydivers were sitting on the floor toward the aft portion of the airplane (they were unsecured). During takeoff, the CG was again well beyond its aft limit. The airplane stalled and crashed. The airplane was destroyed with casualties.

There is a safety factor built into the formula for maximum gross weight. Any airplane can fly when it takes off at a weight greater than maximum gross weight if the runway is long enough and the density altitude is low enough. However, landing is a different matter. All airplanes are built to withstand an occasional hard landing, but what would happen if the hard landing were combined with a substantially overweight airplane? Something would probably break at that time or the structure would be weakened enough to break sometime in the future when everything might seem normal to a pilot unaware of the previous situation. Even more disastrous than an overweight, hard landing is reaching or exceeding the structural integrity of the metal and/or composite design values when maneuvering or when turbulence is encountered. Hidden damage could result, causing an unexpected catastrophic failure at some future time.

If an airplane is certificated with a maximum gross weight of 6,000 pounds (its weight on the ground) and is rolled into a 60° bank, the forces exerted make it feel as if it weighed 12,000 pounds. At its maximum certificated gross weight, there is no problem because the aircraft is operated within its certificated maneuvering loads. But loaded to 8,000 pounds with a 60° bank or an abrupt pull-up, it suddenly weighs 16,000 pounds and might not be able to perform! Even if it could, there would probably be internal stress damage that would show up on future flights.

For more information, see "Aircraft Weight and Balance Definitions" by Sarina Houston on the website: https://www.thebalancecareers.com/aircraft-weight-and-balance-terms-282771.

Unit 5 Navigation Systems I

Lead-in

Air navigation is the process of piloting an aircraft from one geographic position to another while monitoring one's position as the flight progresses. It introduces the need tor planning, which includes plotting the course on an aeronautical chart, selecting checkpoints, measuring distances, obtaining pertinent weather information, and computing flight time, headings, and fuel requirements.

This unit will cover:
- Navigational aids (NAV AIDs)
- Radio waves
- Antenna
- Non-directional Radio Beacon (NDB)
- Automatic Direction Finder (ADF)

Part 1 Listening and Speaking

New Words & Expressions

Aeroflot *n.* /ˈeərəʊflɒt/		a Russian airline company
		俄罗斯航空公司
climb *v.* /klaɪm/		fly up 爬升
checklist *n.* /ˈtʃeklɪst/		a list of things to do or check
		检查单
restore *v.* /rɪˈstɔːr/		recover 恢复
hold *n.* /həʊld/		in waiting （空中）等待
vicinity *n.* /vəˈsɪnəti/		near place 附近
level off		平飞
transponder code		应答机编码

I. Listen to the passage and fill in the blanks with the words you hear.

On February 6, 2021, Aeroflot 2694 from Moscow Sheremetyevo (Russia) to Amsterdam (Netherlands), was _____ out of Moscow when the aircraft _____ at FL280, the crew selected the _____ code for loss of communication. While working the related _____ the crew was able to _____ communication via one of the three available radios, descended the aircraft to FL170 and entered a _____ to burn off fuel and returned to Moscow for a safe landing on Sheremetyevo's runway 24L about 2.5 hours after _____.

The airline confirmed the aircraft returned due to a partial failure of the radio communication system.

II. Filling the blanks with the words from the box.

① curvature	② signal	③ sky
④ frequency	⑤ refracted	⑥ navigation

There are three different types of waves, each with different characteristics. they are ground waves, sky waves and space waves. Ground waves are lower _____ waves that travel close to the surface of the Earth, and will, in fact, follow the _____ of the Earth. The lower the frequency of a wave, the further the _____ will be able to travel. These ground waves travel reliably and predictably along the same route day after day, not being influenced by outside factors. _____ waves are higher frequency waves. They can also travel for long distances, but instead of following the curvature of the Earth, the waves are _____ or bent by the ionosphere and sent back down to Earth. Using sky waves, high frequency radios can send messages across , using only 50 to 100 watts of power. Space waves consist of very high frequency waves, or higher that neither bend nor refract. These waves travel in a straight line, passing through the ionosphere and allow _____ from space.

III. Read the dialogue and Discuss.

C – Air Traffic Controller　　　　　　　　P – Pilots

P: Shanghai Approach, CCA102, our weather radar is inoperative, request weather information.

C: CCA102, there is no weather reported by other aircrafts in your vicinity, you can continue your flight plan.

Discussion

1. What is radio communication system used for?
2. How should the crew respond when encountering radio communication failure?

Part 2 Reading

Text A Intensive Reading

Navigation Systems (1)

Aside from **pilotage** and **dead reckoning**, other forms of **navigation** are also available to pilots through the use of electronic **navigational** aids, or NAV AIDs for short. These systems **transmit signals** to aircraft through **radio waves** and tell pilots where they are and where to go. Before we get into what these systems are and how they work, we first need to review radio waves and **antenna**s. Radio waves are a type of **electro-magnetic radiation**. **Artificially** generated radio waves are used for **fixed** and **mobile communication**, broadcasting, **radar**, computer networks, and, of course, navigation. These artificially generated radio waves are created from antennas. Antennas convert the electric current of a signal into a radio wave, so it can travel through space to a receiving antenna which then **convert**s it back into an **electric current** to be used by a **receiver**. The antennas on an airplane are all different sizes and shapes. This is because they receive or transmit different types of radio waves.

There are three different types of waves, each with different **characteristics**. they are **ground waves**, **sky waves** and **space waves**. Ground waves are lower **frequency** waves that travel close to the surface of the Earth, and will, in fact, follow the **curvature** of the Earth. The lower the frequency of a wave, the further the signal will be able to travel. These ground waves travel **reliably** and **predictably** along the same route day after day, not being influenced by outside factors. Sky waves are higher frequency waves. They can also travel for long distances, but instead of following the curvature of the Earth, the waves are **refracte**d or **bent** by the **ionosphere** and sent back down to Earth. Using sky waves, high frequency radios can send messages across oceans, using only 50 to 100 **watt**s of power. Space waves consist of very high frequency waves, or higher that neither bend nor **refract**.

These waves travel in a straight line, passing through the ionosphere and allow navigation from space.

Most major navigation systems these days operate with signals broadcasting space waves. Objects between the transmitter and receiver may reflect and **block** the signal. This means that the waves have to have a **line-of-sight** between the two, for the signal to be received. For aircraft, the airplane has to be within line-of-sight of the navigational aid in order for the system to work.

One of the oldest types of NAV AIDs still in use today is called the **non-directional radio beacon**, or NDB. And NDB is simply just a ground-based AM (**Amplitude Modulation**) radio transmitter that transmits radio waves in all directions. Because NDBs operate in low to medium **frequency band**, and they are not **subject**ed to the line-of-sight limitations of space waves.

To navigate via NDBs, pilots need to have **install**ed in their aircraft an automatic direction finder, or **ADF**. The face of an ADF contains a **needle** that points to the **relative bearing** of the NDB. The relative bearing is the number of degrees measured **clockwise** between the aircraft's **heading** and the **direction** from which the bearing is taken from. You can use this simple **formula** to calculate the **magnetic bearing** to the station: MH + RB = MB (magnetic heading plus relative bearing equals magnetic bearing). For example, if your airplane is flying a heading of 030 and the ADF is indicating a relative bearing of 120, then that means that the NDB is at a relative bearing of 150 degrees. If you wanted to fly towards the NDB, 150 would be the **initial** heading to turn to. When flying directly toward the NDB, the needle will look like this: pointed straight up at the NDB station. Once crossing over it, the needle will **reverse** direction but still point at the NDB as you fly away from it. It is possible to track both "towards" and "away from" an NDB station. This sounds really easy, right? Just keep the needle straight up and you'll fly right towards the station. Well, in a no wind situation that would probably be just fine. However, most of the time there is in fact wind. If a pilot were to keep the needle straight up on a windy day as they were navigating, they'd be doing a procedure called homing. **Homing** is not a recommended procedure to follow, as you would not be flying in a straight line. Instead of homing, tracking should be used to fly from station to station in a straight line. **Tracking** involves **compensat**ing for the wind by turning slightly into the wind and thereby staying on course. When you are on course and tracking to the station, the airplanes' **wind correction angle** should equal the number of degrees the ADF is deflected from straight up.

NDBs can be spotted on **sectional charts** with this – **magenta** colored symbol. In the **vicinity** of the symbol, you will find a box containing the name of the NDB, frequency, ID and associated **Morse** code. Before you can actually navigate via an NDB, you need to tune and identify the desired station. **Tuning** in a station is pretty easy. Just look up the frequency on your chart and enter it into your receiver. After it's entered, we need to **confirm** that we are receiving the correct station, and that it is **operational**. As part of an NDB **transmission**, they'll send out their ID in Morse code format. We, the pilots, must listen to the Morse code and **verify** that it matches what's printed on our chart. After we've identified the station, we can use it for navigation. However, since there is no **flag** on the instrument to advise us whether or not it is operating properly, we must continue to **monitor** the Morse code for as long as we **intend** on using the station. Luckily, we can turn down the volume, so it's not as **obnoxious**.

New Words & Expressions

pilotage *n.* /ˈpaɪlətɪdʒ/	the guidance of ships or airplanes from place to place 领航
navigation *n.* /ˌnævɪˈɡeɪʃn/	the guidance of ships or airplanes from place to place 航行，航海
navigational *adj.* /ˌnævɪˈɡeɪʃənl/	of or relating to navigation 航行的，航运的
antenna *n.* /ænˈtenə/	an electrical device that sends or receives radio or television signals 天线
artificially *adv.* /ˌɑːrtɪˈfɪʃəli/	artificially induced conditions 人为地
fixed *adj.* /fɪkst/	(of a number) having a fixed and unchanging value 确定的
radar *n.* /ˈreɪdɑːr/	measuring instrument in which the echo of a pulse of microwave radiation is used to detect and locate distant objects 雷达，无线电探测器
convert *v.* /kənˈvɜːrt/	change the nature, purpose, or function of sth 转变
receiver *n.* /rɪˈsiːvər/	set that receives radio or TV signals 无线电接收机
characteristics *n.* /ˌkærəktəˈrɪstɪks/	a prominent aspect of something 特性
frequency *n.* /ˈfriːkwənsi/	the number of occurrences within a given time period 频率
curvature *n.* /ˈkɜːrvətʃər/	the rate of change (at a point) of the angle between a curve and a tangent to the curve

曲率

reliably *adv.* /rɪˈlaɪəbli/   in a faithful manner   可靠地

predictably *adv.* /prɪˈdɪktəbli/   in a predictable manner or to a predictable degree
可预言地

refract *v.* /rɪˈfrækt/   refract a light beam
使折射

bent *adj.* /bent/   altered from an originally straight condition
弯曲的

ionosphere *n.* /aɪˈɒnəsfɪr/   the outer region of the Earth's atmosphere; contains a high concentration of free electrons
电离层

Watt *n.* /wɑːt/   a unit of power equal to 1 joule per second; the power dissipated by a current of 1 ampere flowing across a resistance of 1 ohm
瓦特

reflect *v.* /rɪˈflekt/   to throw or bend back or reflect (from a surface)
反射

block *v.* /blɒk/   render unsuitable for passage
阻止

line-of-sight *n.* /ˌlaɪn əv ˈsaɪt/   an imaginary straight line along which an observer looks
视线

subject *v.* /ˈsʌbdʒɪkt; ˈsʌbdʒekt/   make liable or vulnerable to
(使) 受影响

install *v.* /ɪnˈstɔːl/   set up for use
安装

ADF   automatic direction finder 自动测向器

needle *n.* /ˈniːdl/   a slender pointer for indicating the reading on the scale of a measuring instrument
指针

clockwise *adv.* /ˈklɒkwaɪz/   in the direction that the hands of a clock move
顺时针方向地

heading *n.* /ˈhedɪŋ/   the direction or path along which something moves or along which it lies
航向

direction *n.* /dəˈrekʃn; daɪˈrekʃn/ the spatial relation between sth and the course along which it points or moves
方向

formula *n.* /ˈfɔːrmjələ/ a group of symbols that make a mathematical statement
公式

initial *adj.* /ɪˈnɪʃl/ occurring at the beginning
最初的

reverse *v.* /rɪˈvɜːrs/ change to the contrary
反转

homing *adj.* /ˈhəʊmɪŋ/ orienting or directing homeward or to a destination
归航的

tracking *n.* /ˈtrækɪŋ/ the pursuit (of a person or animal) by following tracks or marks they left behind
追踪

compensate *v.* /ˈkɒmpenseɪt/ make amends for; pay compensation fo
补偿

magenta *adj.* /məˈdʒentə/ of deep purplish red
品红色的

vicinity *n.* /vəˈsɪnəti/ a surrounding or nearby region
邻近

morse *n.* /mɔːs/ a telegraph code in which letters and numbers are represented by strings of dots and dashes (short and long signals)
莫尔斯电码

tune *v.* /tuːn/ adjust for (better) functioning
调整

confirm *v.* /kənˈfɜːrm/ establish or strengthen as with new evidence or facts
确认

operational *adj.* /ˌɑːpəˈreɪʃənl/ pertaining to a process or series of actions for achieving a result
操作的

transmission *n.* /trænzˈmɪʃn, trænsˈmɪʃn/
the act of sending a message; causing a message to be transmitted
传送

verify *v.* /ˈverɪfaɪ/ confirm the truth of
核实

flag *n.* /flæg/	a device, symbol or drawing typically resembling a flag used as a marker （用作标志的）旗状装置
monitor *v.* /ˈmɒnɪtər/	keep tabs on; keep an eye on; keep under surveillance 监控
intend *v.* /ɪnˈtend/	have in mind as a purpose 打算
obnoxious *adj.* /əbˈnɑːkʃəs/	causing disapproval or protest 可憎的
dead reckoning	航位推测法
transmit signal	传输信号
radio waves	无线电波
electro-magnetic radiation	电磁辐射
mobile communication	移动式通信
electric current	电流
ground waves	地面电波
sky waves	天空电波
space waves	太空波
non-directional radio beacon	无方向性的无线电信标
amplitude modulation	调幅，波幅调制
frequency band	频段
relative bearing	相对方位,舷角
magnetic bearing	磁向位
wind correction angle	风修正角
sectional chart	截面图

Exercises

I. Read the above passage and answer the following questions.

1. What forms of navigation systems are there?
2. What are radio waves? Where are they from?
3. What are the differences among ground waves, sky waves and space waves?
4. Why is ADF installed in NDB navigation?

II. Multiple choice (one or more answers).

1. What is the function of NAV AIDs?
    A. These systems transfer signals to aircraft and tell pilots where to go.
    B. These systems change signal to aircraft through radio wave.
    C. These systems tell pilots how to control the aircraft.
    D. These systems review radio waves and antennas and tell pilot where they're.

2. What do you know about antennas?

  A. The artificially generated radio waves are created from antennas.

  B. Antennas convert the electric current of a signal into a radio wave and vice versa.

  C. The antennas on an airplane are all different sizes and shapes.

  D. The antennas on an airplane receive or transmit different types of radio waves.

3. Which statement about the different types of waves is true?

  A. The lower the frequency of a wave, the further the signal will be able to travel.

  B. Ground wave are higher frequency waves that travel close to the surface of the Earth.

  C. Sky waves are higher than space waves.

  D. Space waves are not subjected to the line-of-sight.

4. How can navigation systems that operate with space signals work for aircraft?

  A. The airplane has to be within line-of-sight of the navigational aid.

  B. The airplane has to reflect the signal.

  C. The airplane has to operate in low to medium frequency band.

  D. The airplane has to be within line-of-limitations of space waves.

5. What should pilots do while navigating via NDB?

  A. Pilots need to have installed in their aircraft an automatic direction finder.

  B. Pilots don't need to calculate the magnetic bearing to station.

  C. Parents just need to keep the needle straight up all the way.

  D. Pilots need to track where there is wind.

Text B Extensive Reading

Navigation Systems (2)

There are four different classes of NDBs, they all operate on the same principles, but the different classes **contrast** how far away their signal can be reached. The weakest of all NDBs is the **compass locator**. This low powered NDB uses less than 25 watts of power, giving it a range of only 15 **nautical miles.** The other three classes are **label**ed as medium-high, high and high-high, with each offering progressively larger **range**s.

To receive a signal from an NDB, the aircraft's ADF is able to determine the relative bearing from the aircraft to the NDB station. This is accomplished through the use of two antennas **onboard** the aircraft: one being the **loop antenna**, the other being the **sense antenna**. The loop antenna is a directional antenna, containing two or more **stationary** loops of wire. Looking at just one loop, if radio waves hit the loop in any direction other than directly **perpendicular**, a **voltage** will be induced over the antenna. By using **multiple** loops **oriented** in different **heading**s,

contrast *v.* /ˈkɒntrɑːst/
有明显的差异

compass locator 罗 盘 定
位器

nautical miles 海 里（ 合
1.852 公里）

label *v.* /ˈleɪbl/ 把……归类

range *n.* /reɪndʒ/ 范围

onboard *adv.* /ˈɑːnbɔːrd/
在飞机上

loop antenna 回路天线

sense antenna 辨向天线

stationary *adj.* /ˈsteɪʃəneri/
固定的

perpendicular *adj.*
/ˌpɜːrpənˈdɪkjələr/ 垂直的

the system can **deduce** down to two possible headings that the signal is coming from, both 180 degrees apart. To remove this **ambiguity**, the sense antenna is also used. This antenna, which is more or less just a straight wire, looks at the **electrical field** of the signal, receiving an **identical** signal from all directions. Looking at the **phase** of the signal and not the amplitude, the ADF receiver compares the sense antenna signal with the loop antenna, and is able to remove the ambiguity, and **deduce**s the relative bearing of the NDB station.

Now, before you start relying on NDBs for navigation, you should be **aware** of its limitations. The first of many errors is called the thunderstorm effect. During a **thunderstorm**, the ADF needle will be **temporarily** deflected towards the **lightning** strikes, instead of the NDB. Next is the night effect. Where NDB signals can be refracted by the ionosphere and returned as sky waves, this effect is the largest during the **dawn** and **dusk** hours. This can cause **interference** with distant NDB stations. Mountains can also have an effect on the NDB signal, as they can reflect the NDB signal. Finally, there's the **coastal** effect. As the airplane is flying across a coastline, the ADF needle will bend slightly towards the coastline when crossing it at an angle. All of these errors result in **erroneous** bearing information, which affects the ADF needle. Since the pilot has to **monitor** the NDB Morse code, hearing any **static** on that frequency along with the ADF needle acting **erratically** are two indicators that there may be an error at what you are receiving.

While NDBs are a dying technology, our next NAV AID is still very much alive and much more common in the National Airspace System. This type of NAV AID is called a Very High Frequency **Omnidirectional** Range, also known as a VOR. The typical VOR is usually white and **resembles** a big **bowling pin**. However, other types of VOR exist that look much different. Since VORs operate within the frequency band of 108 to 117.95 megahertz, they fall into the space wave **spectrum**. This does allow for relatively interference-free navigation. However, **reception** is limited to line-of-sight, which prevents a pilot from receiving a signal when at low altitudes or in mountainous terrain.

VORs are **oriented** to **magnetic** north and transmit **radial** information outward in every direction, similar to spokes on a bicycle. Technically speaking, there are an infinite number of radials being broadcast out, but for simplicity, it is said that only

voltage *n.* /ˈvəʊltɪdʒ/ 电压
multiple *adj.* /ˈmʌltɪpl/ 多重的
oriented *v.* /ˈɔːrientɪd/ 使朝向
heading *n.* /ˈhedɪŋ/ 方向
deduce *v.* /dɪˈduːs/ 推断
ambiguity *n.* /ˌæmbɪˈgjuːəti/ 不明确
electrical field 电场
identical *n.* /aɪˈdentɪkl/ 完全相同的
phase *n.* /feɪz/ 阶段，相位
aware *adj.* /əˈwər/ 意识到的
thunderstorm *n.* /ˈθʌndərstɔːrm/ 雷暴
temporarily *adv.* /ˌtempəˈrerəli/ 临时地
lightning *n.* /ˈlaɪtnɪŋ/ 闪电
dawn *n.* /dɔːn/ 黎明
dusk *n.* /dʌsk/ 黄昏
interference *n.* /ˌɪntərˈfɪrəns/ 干扰
coastal *adj.* /ˈkəʊstl/ 沿海的
erroneous *adj.* /ɪˈrəʊniəs/ 错误的
monitor *v.* /ˈmɒnɪtər/ 监控
static *n.* /ˈstætɪk/ 静电干扰
erratically *adv.* /ɪˈrætɪkli/ 不定地
omnidirectional *n.* /ˌɑːmnɪdəˈrekʃənl, ˌɑːmnɪdaɪˈrekʃənl/ 全方向天线
resemble *v.* /rɪˈzembl/ 类似
bowling pin 保龄球瓶
spectrum *n.* /ˈspektrəm/ 光谱，频谱
reception *n.* /rɪˈsepʃn/ 感受

360 radials are used.

In the airplane, the VOR indicator consists of three vital parts: the **course deflection indicator**, or CDI, the TO/FROM indicator, and the **omnibearing selector**, or OBS. The OBS **knob** is used to choose the course or radial that you'd like to reference. The TO/FROM flag will tell you whether the radial selected will take you toward the VOR or away from it. And finally, the CDI tells you how far off you are from the center of the course in degrees.

Now, you may be asking yourself: self, what's the difference between a radial and a course? Well, they are really the same thing. However, when flying, radials are directed away from the station whereas courses are directed toward the station. So, when you fly away from a station, you want to follow a radial, when you fly toward a station, you want to follow a course. The reason for the **distinction** is because the VOR indicator that is an **old-school fashion** and does not know what the aircraft's heading is. Let's say you are directly south of the VOR, flying **northbound**, you could **dial** in either the 180 degrees radial "FROM" or the 360 degrees course "TO" into the OBS, and both will tell you that you are on course. However, if you start **drift**ing to the west, you'd get two different indications on your instrument. The instance, where you have 360 degrees "TO" selected, will indicate that you are left of center. However, if you had 180 degrees "FROM" selected, the instrument will actually indicate that you are right of center. This type of situation is called **reverse sensing**. If you were not aware that, you had mistakenly entered the **reciprocal radial** into the instrument. The more you tried to correct toward the center, the further off course you would actually get.

Modern-day **avionics**, like the Garmin 1000, use a horizontal situation indicator, or HIS, instead of the old **standalone** instruments. HSI is **merge** your heading and navigation into one instrument, and because of that, does not **succumb** to reverse sensing, at least with VORs. However, a pilot navigating with an HIS should still always dial in the appropriate radial or course, because, if, by chance, there was a failure with the instrument and it no longer **sync**ed with your heading, you would not want to suddenly **encounter** reverse sensing.

Sense of VORs radials **emit** out like spokes on a bicycle, the closer the pilot flies to the VOR, the more sensitive the instrument gets. Let's again say that we are south of the VOR on the 180 radial, flying northbound, we set the OBS to 360 degrees and we

magnetic *adj.* /mæɡˈnetɪk/
地磁的
radial *n.* /ˈreɪdiəl/ 辐射道路
course deflection indicator
航向偏差指示器
omnibearing selector 无
线电定向标选择器
knob *n.* /nɒb/ 旋钮
distinction *n.* /dɪˈstɪŋkʃn/
差别

old-school fashion 老派的
东西或做法
northbound *adj.*
/ˈnɔːrθbaʊnd/ 北行的
dial *v.* /ˈdaɪəl/ 拨号
drift *v.* /drɪft/ 漂泊

reverse sensing 反向感应
reciprocal radial 反向径向

avionics *n.* /ˌeɪviˈɒnɪks/
航空电子设备
standalone *adj.*
/ˈstændəˌləʊn/ 独立运行的
merge *v.* /mɜːrdʒ/ 融合
succumb *v.* /səˈkʌm/ 屈服
sync *v.* /sɪŋk/ 使同步

encounter *v.* /ɪnˈkaʊntər/
遭遇
emit *v.* /iˈmɪt/ 发出，散发
（气体，辐射物）

get a "TO" indication. The closer we get to the station, the more sensitive the needle gets. But the indication will continue to show "TO". As you pass over the top of the VOR, you enter a zone called the cone of confusion. The cone of confusion is the area above the VOR, where the airplane does not get a clear signal. The TO/FROM indicator will go to the "off" position, because the receiver can't quite tell where you are. As you fly away from the VOR, the receiver gets the signal again and the flag **flip**s to a "FROM" indication. Now, you can track the 360 degree radial from the VOR and continue flying northbound, tracking the radial away from the station.

flip *v.* /flɪp/ 翻转

If at any time you want to figure out where you are in relation to a VOR, all you need to do is find what radial you are on. That means that on your indicator you need a centered CDI (Course Deviation Indicator) needle and a "FROM" flag, simply keep rotating the OBS knob until the CDI centers. If by chance it's centered with a "TO" flag, you are on the reciprocal radial, you need to rotate the OBS 180 degrees left or right. It will center once again, this time with a "FROM" flag.

So, by using one VOR, you would know where you are in relation to that VOR. However, you don't know at what point you are along that specific radial. For that, you'd need either distance measuring equipment, or DME, or a second VOR. The DME will tell you how far from the VOR you are, **pinpoint**ing your location. 2 VORs can accomplish the same thing through a process called **triangulation**. To **triangulate** your position, pick 2 VORs that are near you and tune in their **respective** frequencies. Now, simply center both needles with "FROM" flags to find the radials. Use a **sectional chart** to draw the radials out, the two radials should **intersect**, indicating your current location.

pinpoint *v.* /ˈpɪnpɔɪnt/ 精确地找到
triangulation *n.* /traɪˌæŋɡjʊˈleɪʃn/ 三角测量
triangulate /traɪˈæŋɡjəˈlɪt/ *v.* 把……分成三角形
respective *adj.* /rɪˈspektɪv/ 分别的
intersect *v.* /ˌɪntərˈsekt/ 相交
regulation *n.* /ˌreɡjʊˈleɪʃn/ 规则

Before any flight that you intend on using your VOR receiver, you should make sure it works. FAA **regulations** require you to check your VOR equipment every 30 days for **IFR** (Instrument Flight Rules) operations, it's not required for **VFR** (Visual Flight Rules) operations but it's a good idea to test it anyway. When pilots perform a VOR check, a record of it is kept in the airplane. This **log** contains the date of the check, the location of the check, any bearing errors encountered during the check, and finally the pilot's signature.

IFR *n.* 仪表飞行规则
VFR *n.* 目视飞行规则
log *n.* /lɒɡ/（航海或飞行）日志
accomplish *v.* /əˈkɒmplɪʃ/ 完成
cone of confusion 锥形干扰区
sectional chart 截面图

Notes

1. pilotage 领航

Pilotage is a term that refers to the sole use of visual ground references. The pilot identifies landmarks, such as rivers, towns, airports, and buildings and navigates among them. The trouble with pilotage is that, often, references aren't easily seen and can't be easily identified in low visibility conditions or if the pilot gets off track even slightly.

2. dead reckoning 航位推算法

Dead reckoning involves the use of visual checkpoints along with time and distance calculations. The pilot chooses checkpoints that are easily seen from the air and also identified on the map and then calculates the time it will take to fly from one point to the next based on distance, airspeed, and wind calculations. A flight computer aids pilots in computing the time and distance calculations and the pilot typically uses a flight planning log to keep track of the calculations during flight.

Exercise

I. Answer the following questions.

1. How do the ADF determine the relative bearing from the aircraft to the NDB station?

2. What are the limitations of NDBs?

3. What does the pilot have to monitor in case of possible errors?

4. What are the advantages and disadvantage of VOR?

5. Why should reverse sensing be avoided?

6. Why should you use DME or 2 VORs?

II. Multiple choice (one or more answers).

1.What is the order of the four different types of NDB based on the range of their signal?

    A. High–low, high, medium–high, high–high.

    B. Highest, higher, medium–high, high.

    C. Low, high , medium–high, high–high.

    D. Low, high, medium–high, medium–low.

2. What are the conditions under which voltage is sensed on the antenna?

    A. Sky waves hit the loop in any direction other than directly perpendicular.

    B. Radio waves hit the loop in any direction other than directly perpendicular.

    C. When flying, radials are directed away from the station whereas courses are directed toward the station.

    D. The antennas on the aircraft are blocked from receiving the signal by high terrain.

3. Which statement about a radial and a course is true?

    A.When flying, radials are directed toward from the station whereas courses are directed toward the station.

  B. Radials are directed away from the station whereas courses are directed away from the station.

  C. When flying, courses are directed away from the station whereas radials are directed toward the station.

  D. When flying, radials are directed away from the station whereas courses are directed toward the station.

4. How do you know the distance between you and VOR?

  A. The DME will show.

  B. The DME will tell how far you are from the VOR by pinpointing your location.

  C. 2 VORS will tell how far you are from the VOR by triangulation.

  D. 3 VORS will show where you are.

Part 3 Translation

1. Most major navigation systems these days operate with signals broadcasting space waves. Objects between the transmitter and receiver may reflect and block the signal. This means that the waves have to have a line-of-sight between the two, for the signal to be received. For aircraft, the airplane has to be within line-of-sight of the navigational aid in order for the system to work.

2. To navigate via NDBs, pilots need to have installed in their aircraft an automatic direction finder, or ADF. To receive a signal from an NDB, the aircraft's ADF is able to determine the relative bearing from the aircraft to the NDB station. This is accomplished through the use of two antennas onboard the aircraft: one being the loop antenna, the other being the sense antenna.

3. While NDBs are a dying technology, our next NAV AID is still very much alive and much more common in the National Airspace System. This type of NAV AID is called a Very High Frequency Omnidirectional Range, also known as a VOR.

4. 为了获得与量距仪站的距离，你的飞机的测距仪（DME）接收器首先会向量距仪站发送一个信号。量距仪站回应。然后，飞机的接收器计算完成信号传输所需的时间，并将其转换为距离。

5. 现代航空电子设备，如佳明 1000 综合声讯控制系统（Garmin 1000），使用水平态势指示器（HIS），而不是旧的独立仪器。HIS 将航向和导航合并到一个仪器中，因此，在甚高频全向信标（VOR）协助下，其不盲从于反向传感。

6. 在任何打算使用 VOR 接收机的执飞航班之前，应该确保它能正常工作。美国联邦航空局规定，按照仪表飞行规则（IFR）飞行的飞机，VOR 设备必须每 30 天检查一次，而按照目视飞行规则（VFR）飞行的飞机不需要做此检查，但无论如何，经常检测 VOR 设备都是有益无害的。

Part 4 Supplementary Reading

How Pilots Use Air Navigation to Fly

Air navigation is accomplished by various methods. The method or system that a pilot uses for navigating through today's airspace system will depend on the type of flight that will occur (VFR or IFR), which navigation systems are installed on the aircraft, and which navigation systems are available in a certain area.

The following are some of the key tools and techniques pilots use for navigation.

Radio Navigation Methods for Aircraft

With aircraft equipped with radio navigation aids (NAVAIDS), pilots can navigate more accurately than with dead reckoning alone. Radio NAVAIDS come in handy in low visibility conditions and act as a suitable backup method for general aviation pilots that prefer dead reckoning. They are also more precise. Instead of flying from checkpoint to checkpoint, pilots can fly a straight line to a "fix" or an airport. Specific radio NAVAIDS are also required for IFR operations.

There are different types of radio NAVAIDS used in aviation:

Automatic Direction Finder and Nondirectional Radio Beacon

The most elementary form of radio navigation is the ADF/NDB pair. An NDB is a nondirectional radio beacon that is stationed on the ground and emits an electrical signal in all directions. If an aircraft is equipped with an automatic direction finder (ADF), it will display the aircraft's position in relation to the NDB station on the ground.

The ADF instrument is basically an arrow pointer placed over a compass card-type display. The arrow always points in the direction of the NDB station, which means that if the pilot points the aircraft in the direction of the arrow in a no-wind situation, they will fly directly to the station. The ADF/NDB is an outdated NAVAID, and it's a system prone to errors.

Since its range is line-of-sight, a pilot can get erroneous readings while flying in mountainous terrain or too far from the station. The system is also subject to electrical interference and can only accommodate limited aircraft at once. Many are being decommissioned as GPS becomes the primary navigation source.

VHF Omnidirectional Range (VOR)

Next to GPS, the VOR system is probably the most commonly used NAVAIDS in the world. VOR, short for VHF Omnidirectional Range, is a radio-based NAVAID that operates in the very-high-frequency range. VOR stations are located on the ground and transmit two signals—one continuous 360-degree reference signal and another sweeping directional signal.

The aircraft instrument (OBI) interprets the phase difference between the two signals and displays the results as a radial on the OBI (omni-bearing indicator) or HSI (horizontal situation indicator), depending on which instrument the aircraft uses. In its most basic form, the OBI or HSI depicts which radial from the station the aircraft is located on and whether the aircraft is flying toward or away from the station.

VORs are more accurate than NDBs and are less prone to errors, although the reception is still susceptible to line-of-sight only.

Distance Measuring Equipment (DME)

DME is one of the most simple and valuable NAVAIDS to date. It's a basic method using a transponder in the aircraft to determine the time it takes for a signal to travel to and from a

DME station. DME transmits on UHF frequencies and computes slant-range distance. The transponder in the aircraft displays the distance in tenths of a nautical mile.

A single DME station can handle up to 100 aircraft at one time, and they usually co-exist with VOR ground stations.

Instrument Landing System (ILS)

ILS is an instrument approach system used to guide aircraft down to the runway from the approach phase of flight. It uses both horizontal and vertical radio signals emitted from a point along the runway. These signals intercept to give the pilot precise location information in the form of a glideslope—a constant-angle, stabilized descent path all the way down to the approach end of the runway. ILS systems are widely in use today as one of the most accurate approach systems available.

GPS Navigation

The global positioning system has become the most valuable method of navigation in the modern aviation world. GPS has proven to be tremendously reliable and precise and is probably the most common NAVAID in use today.

The global positioning system uses 24 US Department of Defense satellites to provide precise location data, such as aircraft position, track, and speed to pilots. The GPS system uses triangulation to determine the aircraft's exact position over the earth. To be accurate, a GPS system must have the ability to gather data from at least three satellites for 2-D positioning, and four satellites for 3-D positioning.

GPS has become a preferred method of navigating due to the accuracy and ease of use. Though there are errors associated with GPS, they are rare. GPS systems can be used anywhere in the world, even in mountainous terrain, and they aren't prone to the errors of radio NAVAIDS, such as line-of-sight and electrical interference.

How Pilots Use NAVAIDS

Pilots will fly under visual flight rules (VFR) or instrument flight rules (IFR), depending on the weather conditions. During visual meteorological conditions (VMC), a pilot might fly by using pilotage and dead reckoning alone, or they might use radio navigation or GPS navigation techniques. Basic navigation is taught in the early stages of flight training.

In instrument meteorological conditions (IMC) or while flying IFR, a pilot will need to rely on cockpit instruments, such as a VOR or GPS system. Because flying in the clouds and navigating with these instruments can be tricky, a pilot must earn an FAA Instrument Rating to fly in IMC conditions legally.

Currently, the FAA is emphasizing new training for general aviation pilots in technologically advanced aircraft (TAA). TAA are aircraft that have advanced highly technical systems onboard, such as GPS. Even light sport aircraft are coming out of the factory with advanced equipment these days. It can be confusing and dangerous for a pilot to attempt to use these modern cockpit systems in-flight without additional training, and current FAA training standards haven't kept up with this issue.

The FAA's updated FITS program finally addressed the issue, although the program is still voluntary.

Lead-in

This unit will cover:
- Very High Frequency Omnidirectional Range (VOR)
- Distance Measuring Equipment (DME)
- Global Positioning System (GPS)
- Wide Area Augmentation System (WAAS)

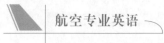

Part 1 Listening and Speaking

New Words & Expressions

visibility *n.* /ˌvɪzəˈbɪləti/	how far and how clear one can see 能见度	
clear *v.* /klɪr/	get approval for sth.　批准放行	
final *n.* /ˈfaɪnl/	（进近）第五边	
threshold *n.* /ˈθreʃhəʊld/	entrance of the runway 跑道（着陆）入口	
localizer *n.* /ˈləʊkəˌlaɪzər/	component of an ILS　航向道	
align *v.* /əˈlaɪn/	make sth in a row　使对齐	
detect *v.* /dɪˈtekt/	find　发现	
query *v.* /ˈkwɪri/	ask about　询问	
heading *n.* /ˈhedɪŋ/	direction　航向	
intercept *v.* /ˌɪntərˈsept/	stop and cut sth off　拦截，切入	
MSL	Mean Sea Level　平均海平面	

I. Listen to the passage and fill in the blanks with the words you hear.

On January 9, 2021, Egypt Air 714 was on approach to Cairo in low _____, low visibility operations were in progress, and was _____ for the ILS approach to runway 05L. While turning _____ over the Nile River, about 10.5 nm from the _____ of runway 05L, the aircraft went through the _____ runway 05L and _____ with runway 05C. ATC _____ the false line up and _____ the crew that they were cleared for 05L, not 05C. Subsequently ATC instructed the crew to turn left _____ 360 and _____ localizer 05L. The crew intercepted localizer 05L and descended reaching 400 feet MSL about 2.6 nm before the threshold runway 05L, where the aircraft should have been at 825 feet MSL. The crew subsequently went around, positioned for another approach to runway 05L and landed safely about 15 minutes after the go around.

II. Filling the blanks with the words from the box.

① terminal	② volumes	③ nautical	④ altitude
⑤ airways	⑥ line-of-sight	⑦ coverage	⑧ layers

Just like with NDBs, there are different classes of VOR, each with their own service _____. VORs have a power output necessary to provide _____ within their assigned operational service volume. There are three service volumes that a VOR can have: terminal, low, and high. A _____ VOR is one usually located on an airport and is a lower powered VOR. It has a range of 25 _____ miles and reaches an altitude of 12,000 feet AGL (Above Ground Level). A low VOR is more powerful and has a range of 40 nautical miles and reaches an _____ of 18,000 feet AGL. And finally, there's the high VOR. This VOR is used to build the high-altitude _____ that exist from 18,000 feet and above. This VOR is higher powered and has many _____ to it. The first is from the surface up to 14,500 feet AGL, and has a range of 40 nautical miles. The next layer is from 14,500 feet to 18,000 feet and a range of 100 miles. Above that, is a layer from 18,000 feet to 45,000 feet, and a range of 130 miles. And finally, the top layer goes from 45,000 feet to 60,000 feet, and has a range of 100 miles. The rounded shape at the bottom of these service volumes is depicting the _____ characteristics of the VOR.

III. Read the dialogue and Discuss.

C – Air Traffic Controller   P – Pilots

C: Fastjet 807, I say again, Fastjet 807,do you read?

P: Approach, Fastjet 807. Reading you five.

C: Fastjet 807, did you read my previous transmissions?

P: Negative, sir. Princeton has just put us over to you.

C: Fastjet 807, that's OK. Your present heading please?

P: Heading 255, Fastjet 807.

C: Fastjet 807, make it 270.

P: Heading 270, Fastjet 807.

Discussion

1. If the aircraft performs wrong ILS, what should be done?
2. What does "read you five" mean?

Part 2 Reading

Text A Intensive Reading

Navigation Systems (3)

Keep in mind that before you actually rely on a particular VOR for navigation, you need to make sure you have the correct VOR tuned in. This is **accomplish**ed by identifying the VOR. Just like an NDB, VORs transmit out three-letter **identification**s in Morse code. After a pilot tunes in a VOR, they need to listen to that VOR's **audio transmission**, and make sure the Morse code **match**es what it's supposed to be. Pilots do not, however, need to know Morse code. All they have to do is match the audio with the **dots and dashes** printed on the sectional chart for that station. Unlike an NDB, pilots do not need to continually monitor the audio of a VOR. Once they have identified the station, they can turn the audio off. If the VOR was **inoperative**, the TO/FROM flag would show "off". As part of another convenience of today's advanced avionics packages, systems like the Garmin 1000, can identify a VOR for you. Once you enter the frequency into your navigational radio, the G1000 displays the VOR's identification next to it. If that ID matches what's on the sectional, you have the right station tuned in. You do not need to listen to the Morse code, unless you want to, of course.

Just like with NDBs, there are different classes of VOR, each with their own **service volume**s. VORs have a power output necessary to provide **coverage** within their **assigned** operational service volume. There are three service volumes that a VOR can have: terminal, low, and high. A terminal VOR is one usually located on an airport and is a lower powered VOR. It has a range of 25 nautical miles and reaches an altitude of 12,000 feet **AGL** (Above Ground Level). A low VOR is more powerful and has a range of 40 nautical miles and reaches an altitude of 18,000 feet AGL. And finally, there's the high VOR. This VOR is used to build the high-altitude **airway**s that exist from 18,000 feet and above. This VOR is higher powered and has many **layer**s to it. The first is from the surface up to 14,500 feet AGL, and has a range of 40 nautical miles. The next layer is from 14,500 feet to 18,000 feet and a range of 100 miles. Above that, is a layer from 18,000 feet to 45,000 feet, and a range of 130 miles. And finally, the top layer goes from 45,000 feet to 60,000 feet, and has a range of 100 miles. The rounded shape at the bottom of these service volumes is **depict**ing the line-of-sight characteristics of the VOR.

The way to tell what kind of service volume a VOR has is to check the chart **supplement** and look next to the VOR name for the symbol in **parentheses**. VORs are shown on a sectional chart with this symbol surrounded by a compass rose. This helps pilots **visualize** and draw out radials for their flight planning. Just like with NDBs, in the vicinity of a VOR is a box of information relating to that station. It includes the same **pertinent** information as an NDB would have. Sometimes, VORs have distance measuring equipment, or **DME**, as well, and will be shown by this symbol. The third and final way of VOR shown on the chart is when it is **co-located** with the **military TACAN** (Tactical Air Navigation System). And this VOR is then called a VORTAC. A TACAN is just the military **equivalent** to a VOR system.

You may also notice certain radials being drawn on the sectional connecting different VORs together. These are known as Victor airways / V airways, and serve as **predefined** low altitude routes that pilots can use to navigate along.

So, now that we've covered all that, let's go over how a VOR actually works. Remember that a radio wave looks like a **sine curve**. If you have ever taken a **trigonometry**, you might remember that the sine-wave has some key points: 0 degrees, 90 degrees, 180 degrees, 270 degrees, and it starts over again at 360 degrees. The VOR uses this concept to operate.

The VOR emits two signals: a **reference phase** and a **variable phase**. The reference phase **emit**s outwards in all directions **simultaneously**. The variable phase emits outward in a **rotating** fashion, similar to a lighthouse. This would be a fast lighthouse rotating around 30 times a second. Older VORs were mechanically rotated. Now, they are scanned electronically to achieve the same result with no moving parts. As the variable phase signal rotates around the VOR, the signal will become phase shifted from the reference phase. The two signals are **detect**ed by the aircraft's VOR receiver, and then compared to determine the **phase angle** between them. So, at 90 degrees, the two waves are phase shifted 90 degrees apart. The phase angle is equal to the direction from the station to the airplane. So, this airplane is on the 090 degree radial from the VOR. The same idea is true at the 180 radial and the 270 degree radial. The two signals are phase shifted, and the airplane receivers can detect this difference.

As **previously** mentioned, one of the ways pilots get distance information is through something called distance measuring equipment, or **DME**. This system operates on frequencies in the **UHF** (Ultra High Frequency) **spectrum**, between 962 **megahertz** and 1,213 megahertz. Similar to other systems, DME emits a Morse code ID every 30 seconds to indicate that it is operating correctly. The range of DME is 199 nautical miles, but it only serves the closest 100 aircrafts. They are also subject to line-of-sight restrictions, just like a VOR.

To obtain distance from a station, your aircraft's DME receiver first transmits a signal to the station. The station then replies back. The aircraft's receiver then measures the time it took to complete the trip, and converts that into distance. Because of that, DME gives the pilot something called a **slant range distance**. This means that the distance shown is actually going to be the exact distance to the DME station, not the distance across the ground. So, an airplane at 3,000 feet might get a DME distance of 0.6 nautical miles, when/while in fact the plane is 0.5 nautical miles away from the DME. The effect worsens at higher altitudes. In fact, if you fly over a VOR at 6,000 feet, even though you were right on top of it, the DME will tell you that you are one nautical mile away from it.

New Words & Expressions

accomplish *v.* /əˈkʌmplɪʃ/	put in effect 完成，实现	
identification *n.* /aɪˌdentɪfɪˈkeɪʃn/	evidence of identity; something that identifies a person or thing 识别信息，身份证明	
audio *n.* /ˈɔːdiəʊ/	the audible part of a transmitted signal 音频	
transmission *n.* /trænzˈmɪʃn; trænsˈmɪʃn/	the act of sending a message; causing a message to be transmitted 传输	
match *v.* /mætʃ/	be compatible, similar or consistent; coincide in their characteristics 匹配，相配	
inoperative *adj.* /ɪnˈɒpərətɪv/	not operating 不起作用，无效	

coverage *n.* /ˈkʌvərɪdʒ/ the extent to which something is covered
覆盖

assigned *adj.* /əˈsaɪnd/ appointed to a post or duty
指定的，已分配的

AGL above ground level 离地高度

airway *n.* /ˈeəweɪ/ a designated route followed by airplanes in flying from one airport to another
航线，空中航线

layer *n.* /ˈleɪə(r); leə(r)/ single thickness of usually some homogeneous substance
层

depict *v.* /dɪˈpɪkt/ give a description of
描述

supplement *n.* /ˈsʌplɪmənt/ textual matter that is added onto a publication; usually at the end
附录

parenthese round brackets 括号

visualize *vt.* /ˈvɪʒuəlaɪz/ imagine; conceive of; see in one›s mind round brackets
形象化，可视化

pertinent *adj.* /ˈpɜːtɪnənt/ having precise or logical relevance to the matter at hand
相关的

co-located *adj.* /ˌsiːˈəʊ-ləʊˈkeɪtɪd/ 同时使用

military *adj.* /ˈmɪlətri/ of or relating to the study of the principles of warfare
军事的

TACAN tatical air navigation
战术空中导航设备 (简称塔康)

equivalent *n.* /ɪˈkwɪvələnt/ a person or thing equal to another in value or measure or force or effect or significance etc
对等的事物

predefined *adj.* /ˌpriːdɪˈfaɪnd/ be defined before
预先定义的，预先确定的

trigonometry *n.* /ˌtrɪɡəˈnɒmətri/ the mathematics of triangles and trigonometric functions
三角学

emit *v.* /iˈmɪt/ give off, send forth, or discharge; as of light, heat, or radiation, vapor, etc.
发出，发射

simultaneously *adv.* /ˌsɪmlˈteɪniəsli/	at the same instant 同时地
rotating *adj.* /rəʊˈteɪtɪŋ/	turn on or around an axis or a center 旋转的
detect *vt.* /dɪˈtekt/	discover or determine the existence, presence, or fact of 检测，探测
previously *adv.* /ˈpriːviəsli/	at an earlier time or formerly 以前
DME	distance measuring equipment 测距机
UHF	ultra high frequency 超高频
spectrum *n.* /ˈspektrəm/	an ordered array of the components of an emission or wave 频谱
megahertz *n.* /ˈmegəhɜːts/	one million periods per second 兆赫
dots and dashes	莫尔斯符号，点和线
service volume	服务量，服务流量
sine curve	正弦曲线
reference phase	参考相位
variable phase	可变相位
phase angle	相位角
slant range distance	倾斜距离

Notes

Garmin 1000 佳明 1000

The Garmin G1000 is an integrated flight instrument system typically composed of two display units, one serving as a primary flight display, and one as a multi-function display. Manufactured by Garmin Aviation, it serves as a replacement for most conventional flight instruments and avionics.

Exercises

I. Answer the following questions.

1. How can you make sure you have the correct VOR tuned in?

2. What kind of VOR can have a range of 40 nautical miles and reach an altitude of 18,000 feet AGL?

3. What helps pilots visualize and draw out radials for their flight planning?

4. What are Victor airways / V airways and what are they for?

5. How does your aircraft's DME obtain distance from a station?

II. Multiple choice (one or more answers).

1. What is not expected to do after pilots tune in a VORs?

    A. They need to listen to that VOR's audio transmission.

    B. They match the audio with the dots and dashes printed on the sectional chart for that station.

    C. They need to continually monitor the audio of a VOR after they have identified the station.

    D. They make sure the Morse code matches what is supposed to be.

2. What kind of VOR can have a range of 25 nautical miles and reach an altitude of 12,000 feet AGL?

    A. A low VOR.

    B. A high VOR.

    C. A terminal VOR.

    D. VORs.

3. Which of the following is true about signals the VOR emits?

    A. The reference phase emits outwards in all directions simultaneously.

    B. The variable phase emits outward in a rotating fashion.

    C. The aircraft's VOR receiver can detect the two signals and determine the phase angle between them.

    D. The phase angle is equal to the direction from the station to the airplane.

4. What do you know about your distance to the station?

    A. DME gives you a slant range distance.

    B. DME gives you the exact distance to the DME station.

    C. DME gives you the distance across the ground.

    D. If DME tells you that you are one nautical mile away from the station, you may be right on top of it.

Text B Extensive Reading

Navigation Systems (4)

The Global Positioning System, the space-based navigation system, can trace its roots back to the 70s when testing began. However, the system became fully **operational** back in 1995. The **GPS** system **consist**s of three **element**s: space, control and user.

The space element consists of a **minimum** of 24 **satellites** and 6 **orbital** planes around the Earth. There are usually closer to 30 GPS satellites in orbit at any time. These satellites are in a **medium** Earth orbit at 10,900 nautical miles above the Earth. At

GPS 全球定位系统

operational *adj.*
/ˌɒpəˈreɪʃənl/ 运作的

consist *vi.* /kənˈsɪst/
由……组成

element *n.* /ˈelɪmənt/部分，
元素

minimum *adj.* /ˈmɪnɪməm/

this distance, they orbit the Earth every 12 hours or twice a day, meaning they are not in **geostationary** orbit, like communication and weather satellites. At this orbit, they travel roughly 7,000 miles per hour, and at an **inclination angle** of 55 degrees from the **equator**. This is important because it allows 5 satellites to remain in view at all times from anywhere on Earth, except at the poles. Each satellite is built to last about 10 years, with replacements built and **launch**ed as needed to keep the system running smoothly. GPS satellites are **power**ed by **solar energy**, but have **backup batteries** onboard to keep them running during periods of **solar eclipse**s. And in addition, small **rocket booster**s are located on each satellite, and are used to keep them flying in the correct orbit. Finally, each satellite contains two or three **atomic clock**s which are the key **component**s to getting your position.

The control element consists of round-based monitoring stations, a master control station and ground antennas around the world. The goal of the control element is to ensure the accuracy of the GPS satellite positions and the accuracy of the atomic clocks onboard. The master control station gets data from monitoring stations **pertain**ing to errors with satellite orbits or clocks, and is able to send data and instructions to the satellites to correct for any errors detected, or to move satellites back to their proper orbits.

The user element consists of the combination of the antennas, receivers and **processor**s in the airplane that receives the signal and **calculate**s your GPS position. There are a wide range of receivers, anywhere from **handheld devices** to **panel mounted** to full flight tech systems, but they all perform the same calculations to give you your position, and each has their own limitations that the pilot must be aware of before using them for navigation purposes.

Each GPS satellite transmits the GPS signal in the **microwave range**. The two primary signals are called the L1 and L2 frequencies. The L1 frequency transmits at 1,575.42 megahertz, and is for **civilian use**, the L2 frequency transmits on 1,227.60 megahertz and is **encrypt**ed for use by the military, although these frequencies are not important to know. Each signal transmits a coarse/acquisition code that contains three parts. First, a **pseudo-random** Code, which is an ID code that identifies the transmitting satellite. This also prevents something called **spoof**ing which in simple terms prevents anyone from interfering with the GPS signal. Second is something called the **ephemeris data** which is

最低限额的，至少

satellite *n.* /ˈsætəlaɪt/ 人造卫星

orbital *adj.* /ˈɔːbɪtl/ 轨道的，环绕的

orbit *n.* /ˈɔːbɪt/ 轨道

medium *adj.* /ˈmiːdiəm/ 中间的，中等的

geostationary *adj.* /ˌdʒiːəʊˈsteɪʃənri/ 与地球旋转同步的

inclination angle 倾斜角，[数]倾角

equator *n.* /ɪˈkweɪtə(r)/ 赤道

launch *v.* /lɔːntʃ/ 发射（导弹、火箭等）

power *v.* /ˈpaʊə(r)/ 提供动力，使……有力量

solar energy 太阳能

backup batteries 备用电池，备用电源，备份电源

solar eclipse [天]日食

rocket booster 火箭起动加速器，火箭助推器

atomic clock 原子钟

component *n.* /kəmˈpəʊnənt/ 组成部分，成分，组件

pertain *v.* /pəˈteɪn/ 关于，有关

processor *n.* /ˈprəʊsesə(r)/ [计]处理器，处理程序

calculate *v.* /ˈkælkjuleɪt/ 计算，预测

handheld devices 手持设备

panel mounted 安装在面板上的

microwave range 微波波段，微波范围

civilian use 民用

encrypt *v.* /ɪnˈkrɪpt/ 把……

describing where each GPS satellite should be in orbit at any given time. And third is the **almanac data**, the current date and time and the status of the satellite, whether healthy or unhealthy. The entire GPS signal takes **approximately** 30 seconds to receive.

The idea of how the GPS works is based on a principle called **pseudoranging**. This is the name for the process that allows us to calculate our distance, not by actually measuring distance but calculating it with a time calculation. This is done using the same formula from dead reckoning, Rate × Time = Distance (Rate times Time equals Distance).

Radio waves travel at the speed of light, which is 186,000 miles per second. The GPS satellite sends a signal to the airplane that has the time the signal was sent. The receiver can compare the time the signal was sent to the time the receiver received the signal. Since we know the speed of the signal and the time it took to get from the satellite to the receiver, we can calculate the distance from the satellite to the airplane. So, the GPS signal that is sent out really is more like "I am satellite X, my position is Y, and this information was sent at time Z".

There is one problem though. This means that we are approximately 10,900 miles away from the satellite in all directions, essentially making a **sphere** that we are on. In order to pinpoint our location, we need to use more satellites. By adding a second satellite there are now two spheres that intersect. This makes a circle of where we could be. A circle contains an infinite number of points, so we need to add a third satellite to get our position. By adding a third satellite, the sphere of our possible locations from the third satellite intersects the circle in two locations: one on Earth and one in space. By **eliminating** the location in space, we know our 2D (two dimensional) location. In order to get a 3D (three dimensional) location, a fourth satellite is necessary to remove any ambiguity in the position. The GPS satellite **constellation** is designed to make at least five satellites in view at all times. And most of the time, there are several more received. If we receive more signals from satellites other than the four necessary to get our 3D position, the GPS receiver will use the additional signals in its position calculation, and give the pilot and even more precise location.

Now that we know our 3D position, we need to check the accuracy of the GPS signals. In order to do this, our GPS receiver

加密
pseudo-random *adj.*
/ˈsuːdəʊ; ˈsjuːdəʊ-ˈrændəm/
[自] 伪随机的
spoof *v.* /spuːf/ 哄骗，行骗
ephemeris data 星历表数据
almanac data 卫星星历
approximately *adv.*
/əˈprɒksɪmətli/
大约，近似地
pseudoranging *n.* 伪距

sphere *n.* /sfɪə(r)/
范围，球体

eliminating *v.* /ɪˈlɪmɪneɪtɪŋ/
消除

constellation *n.*
/ˌkɒnstəˈleɪʃn/
[天] 星座；星群

calculates something called Receiver Autonomous Integrity Monitoring, or **RAIM**. This is the system the receiver uses to **verify** the **usability** of the received GPS signals, which warns the pilot of malfunctions in the navigation system. In order for the receiver to calculate RAIM, we need to receive at least five satellites. If RAIM is not **available**, the pilot will receive a message from the GPS to warn him or her that there may be some error in the GPS position. RAIM **outage**s may occur when there are insufficient number of GPS satellites, or there is an unsuitable satellite geometry, either of which can cause the error in the position to become too large.

Just like every other navigational system, there are several GPS errors that a pilot can experience while flying. The first is anytime there are fewer than 24 operational satellites, which may result in a lack of adequate GPS signal. Next, anytime the antennas on the aircraft are blocked from receiving the signal by high **terrain**, such as in a valley, or any time the aircraft's GPS antenna is shadowed by the aircraft structure, like when the aircraft is **banked**. Some other errors that can occur are **harmonic interference** from VHF (Very High Frequency) transmitting devices, satellite atomic clock inaccuracies, receiver or processor errors, or even a bounced or multipath signal reflected from hard objects. There can also be errors caused by the signal traveling through the ionosphere in **troposphere**, which can cause a delay in the signal. Additionally, sometimes there are satellite data transmission errors, which may cause small position errors or **momentary** loss of the GPS signal. Finally, there was an error called selective availability. This is an error caused by the US Department of Defense that can purposely cause error in GPS signals. This error was **discontinued** on May 1, 2000, but may be **reinstated** at any point in the future that the Department of Defense finds it necessary. So, to sum up, all of these errors, when added together, equal an error of +/-15 meters (plus/minus 15 meter), or roughly 45 feet. When selective availability was turned on, the error was +/- 100 meters (plus/ minus 100 meters), or about 300 feet. The less error we have, the better.

To improve the accuracy, **integrity** and **availability** of GPS signals, something called Wide Area Augmentation System, or **WAAS**, was designed. WAAS worked so well that the location error is a mere 10 feet or so. As the GPS signal reaches Earth, it

RAIM (receiver autonomous integrity monitoring) 接收机完整性自主监测

verify vt. /ˈverɪfaɪ/ 核实，检验

usability n. /ˌjuːzəˈbɪləti/ 可用性

available adj. /əˈveɪləbl/ 可获得的

outage n. /ˈaʊtɪdʒ/ 运 行中断

terrain n. /təˈreɪn/ [地 理] 地形，地势

banked adj. /bæŋkt/ 有坡面的，倾斜的

harmonic interference [电子] 谐波干涉，谐波干扰

troposphere n. /ˈtrɒpəsfɪə(r)/ 对流层

momentary adj. /ˈməʊməntri/ 瞬间的，短暂的

discontinued adj. /ˌdɪskənˈtɪnjuːd/ 停止使用的，停止的

reinstated v. /ˌriːɪnˈsteitid/ 使恢复（reinstate 的过去分词）

integrity n. /ɪnˈtegrəti/ 完整性

availability n. /əˌveɪləˈbɪləti/ 可用性，有效性

WAAS（Wide Area Augmentation System）广域扩增系统

is received and monitored by ground-based wide area reference stations. These stations monitor the GPS signal and **relay** the data to a wide area **master station**. At the master station, a correction to the GPS signal is computed, a correction message is prepared and **uplinke**d to one of the geostationary WASS satellites via a ground uplink, and then broadcast on the same L1 frequency as the regular GPS signal. Any GPS receiver that is also WAAS capable will be able to receive the correction message. The receiver will then apply this correction into its GPS position calculation, and display to pilots, and even more accurate position. The WAAS satellites are in an **ideal** position for their geostationary orbit, allowing them to cover a large portion of the Earth. To take full advantage of their location, the WAAS satellites will also **act as** regular GPS satellites.

relay v. /ˈriːleɪ; rɪˈleɪ/接转, 转送

master station[计][通信] 主站，主控台，总机

uplinke v. /ˈʌplɪŋk/ 向上传输，从地面输送信息

ideal adj. /aɪˈdiːəl/ 理想的

act as 充当

Notes

Ionosphere 电离层

The ionosphere is an abundant layer of electrons and ionized atoms and molecules that stretches from about 48 kilometers (30 miles) above the surface to the edge of space at about 965 km (600 mi), overlapping into the mesosphere and thermosphere. This dynamic region grows and shrinks based on solar conditions and divides further into the sub-regions: D, E and F; based on what wavelength of solar radiation is absorbed. The ionosphere is a critical link in the chain of Sun-Earth interactions. This region is what makes radio communications possible.

Exercises

I. Answer the following questions.

1. What is GPS?
2. what is the user element?
3. What do you know about the GPS signal?
4. When may RAIM outages occur?
5. What are the GPS errors that a pilot can experience while flying?
6. How do WAAS improve the accuracy, integrity and availability of GPS signals?

II. Multiple choice (one or more answers).

1. What is true about GPS satellites?
   A. GPS satellites are powered by solar energy.
   B. There are backup batteries onboard to keep GPS satellites running during periods of solar eclipses.
   C. Small rocket boosters are located on each satellite to keep them flying in the correct

orbit.

    D. Each satellite contains two or three atomic clocks which are the key components to getting your position.

2. What is the function of the control element?

    A. Ensure the accuracy of the north pole positions and the atomic clocks onboard.

    B. Ensure the accuracy of the GPS satellite positions and the atomic clocks onboard.

    C. Ensure the accuracy of the GPS satellite positions and proper orbits.

    D. Ensure the accuracy of the GPS satellite or clock positions and the atomic clocks onboard.

3. What is not true about pseudo-random code?

    A. It can describe where each GPS satellite should be in orbit at any given time.

    B. It can identify the transmitting satellite.

    C. It can prevent anyone from interfering with the GPS signal.

    D. It prevents something called spoofing.

4. In order for the receiver to calculate RAIM, how many satellites should be used?

    A. Two.

    B. Four.

    C. Five.

    D. Six.

5. Which of the following may not cause a GPS error?

    A. Fewer operational satellites than 24.

    B. The signal traveling through the ionosphere in troposphere.

    C. The weather in space being cold and dry.

    D.The antennas on the aircraft blocked from receiving the signal by high terrain.

Part 3 Translation

1. The Global Positioning System, or GPS, is the United States version of a Global Navigation Satellite System – GNSS, the space-based navigation system, can trace its roots back to the 70s when testing began. However, the system became fully operational back in 1995. The GPS system consists of three elements: space, control and user.

    _____

    _____

    _____

2. To improve the accuracy, integrity and availability of GPS signals, something called Wide Area Augmentation System, or WAAS, was designed. WAAS worked so well that the location error is a mere 10 feet or so.

    _____

3. The Instrument Landing System (ILS) provides literal and vertical position data necessary to put the airplane on the runway for approach.

4. 对于新手飞行员，尤其是那些驾驶未配备更先进导航系统的飞行员而言，自动定向台系统是一种适合远距飞行的部件。

5. 如果学过三角函数，你可能会记得正弦波有几个关键点：0 度，90 度，180 度，270 度，然后从 360 度重新开始。

Part 4 Supplementary Reading

BeiDou (Compass) Navigation Satellite System

With a long history and a splendid culture, China is one of the important cradles of early human civilization. Chinese people have used the Big Dipper (BeiDou) for identifying directions, ever since ancient times. They invented the world's first navigation device "Compass", which was a great contribution to the development of world civilization and become a bridge of friendship between China and foreign countries. The ancient compass has become the glory of our history, while BeiDou system will become a new contribution of China to mankind.

BeiDou (Compass) Navigation Satellite System is a global navigation satellite system, which is being independently developed and operated by China. It will provide all weather-position velocity and timing services with high accuracy and reliability with users anywhere anytime. It has advantages of quick-positioning location-reporting and information exchange. The development of BeiDou system can be divided into three phases.

Phase I:

In early 1980s, China began to actively study the Navigation Satellite Systems based on dual satellite active positioning. The construction of the demonstration system started in 1994. The system began to provide services in 2000, which made China the third nation in position of an independent navigation satellite system following the U.S and Russia. The Demonstration system consists of three GEO satellites and corresponding ground systems. The cost-efficient system was quickly put into use. It has been providing positioning short message telecommunication and timing services for users in China and nearby areas.

Phase II:

To provide better services for national economic development and to meet increasing demand for satellite navigation. China started to build BeiDou Navigation Satellite System based on passive positioning in 2004. Around 2012 the system will cover the Asia-Pacific region with more than 10 satellites. Currently, several satellites have been successfully launched. And the project is advancing on schedule.

Phase III:

Around 2020 BeiDou system will be completed, and provide global coverage with 35 satellites, offering state-of-the-art services. Worldwide users will enjoy two kinds of high quality services, including open and authorized services. The open service will be free of direct charge with an accuracy of 10m for positioning 0.2m/s for velocity and 20ns for timing. The authorized service will have higher accuracy and better reliability and will be available to authorized users.

From the snow cover of plateau to the Gobi desert, from the northern border to the southern territorial seas. Since 2000, BeiDou Navigation Demonstration System has been widely used in many fields, such as mapping, telecommunications, water conservancy, fishery, transportation, forest fire, prevention disaster relief and public safety. Especially in the Beijing Olympic Games, the frozen disaster, Wenchuan earthquake relief and Shanghai World Expo. The demonstration system has played an important role and produce significant economic and social benefits. BeiDou system has been widely used in the field of China maritime fishery. It can provide services for fishery management development, such as vessel position monitoring, emergency rescue, policy issue and vessel in and out management. It can also provide services for fishing vessels, such as navigation and positioning distress signals maritime notices and value attitude information, providing vessel-vessel and vessel-shore short message telecommunications.

BeiDou system has also played an important role in disaster rescue. In 2008, Wenchuan suffered a catastrophic earthquake, in which the ground cables and communication facilities were damaged seriously. Under the emergency of interruption of regular communications and impossible roads, BeiDou system provided an important assurance for command of relief teams and notice of disaster's situation. And it was an advantage for positioning and short message telecommunications.

As an important member of ICG, China has participated in all previous assemblies and private forums, widely communicated with authorities of relevant countries regions and international organizations. China will always cooperate with foreign countries. And promote cooperation in frequency coordination, compatibility and interoperability, international standard, time and coordinate reference, scientific research and applicative development between BeiDou system and other satellite navigation systems. China will promote worldwide applications of satellite navigation and make contributions to scientific development and technological progress of this field.

Our own ancestors used the Big Dipper for identifying direction. Today we use BeiDou system for global positioning. As one of the cooperators of global navigation satellite system, BeiDou system will be committed to providing satellite positioning, navigation and timing services with high performance high quality and high reliability for global users, and make important contributions to human society.

SERVE THE WORLD, BENEFIT THE MANKIND. This is BeiDou Navigation Satellite System.

Unit 7　Air Traffic Control

Lead-in

　　Air Traffic Control is the service that provides by ground systems due to the safety of the passengers of the aircrafts. This unit outlines the importance of Air Traffic Control in the real life and discusses the types of the Air Traffic Control, as well as some of the key features of the airport operations and duties of Air Traffic Controller.

Part 1 Listening and Speaking

New Words & Expressions

airfield *n.* /ˈeəfiːld/ — an area of flat ground where military or private planes can take off and land
飞机场

bay *n.* /beɪ/ — a partly enclosed area, inside or outside a building, that is used for a particular purpose
(建筑物内或外辟作特定用途的) 隔区，停机位置

inbound *adj.* /ˈɪnbaʊnd/ — travelling towards a place rather than leaving it
到达的，归航的，进站的，入港的

outbound *adj.* /ˈaʊtbaʊnd/ — travelling from a place rather than arriving in it
向外的，离开某地的，出港的，离场的，出航的

WP — waiting position
等待位置

standby *n.* /ˈstændbaɪ/ — ready to do sth immediately if needed or asked
随时可以投入行动，处于待命状态

hold short of — 等待

I. Filling the blanks with the words from the box.

① returning signals	② permission	③ visual checks	④ control tower
⑤ air traffic controllers	⑥ airfield	⑦ traffic jams	⑧ radars
⑨ landing, taking off and moving around			

With around half million people flying into and out of Britain's Airport each day, there are plenty of things to do to make sure the passengers get to the right flights and planes to take off on time. With so many planes _____ the airport, they need to be controlled to ensure that they keep safe distance from each other and don't cause _____. While cars on the road could be controlled using traffic light, lights are not very practical to control planes. So, that's why the _____ coming. So, how do they keep everything running smoothly? Let's find out.

Firstly, as air traffic controllers need to see all around the _____, they need to be high up, so they are found in the _____. From up here, they can see everything that is going on. Well, normally … Fog and storms can make it hard to get a clear view. Luckily, although the team do make _____, they don't have to rely on their eyes. They have high-tech computers to track the movements of the planes, and _____ too, which can see everything that's moving on the airfield and also in the zone around 7 kilometers in the air above the airport. Radars send waves out. And when those waves hit objects, the _____ give the controllers information about what it is, how fast it is moving, and which direction. In addition, anything moves in the airspace around the airport has to check in with the air traffic controllers. Planes can't move around on the airfield, or in the air above it, without getting _____.

II. Listen to the dialogue and answer the following questions.

C – Air Traffic Controller   P1~5 – Pilots
P1: Ground, good evening, SIA173.
C: SIA173, good evening, taxi on the greens for Bay F33.
P1: On the greens for F33, SIA173.
C: BAW430, the Pacific 737 inbound on WP will be giving way to you.
P2: BAW430, traffic in sight.
C: DLH15, how long before ready for taxi, sir?
P3: Ah, 2 minutes, DLH15.
C: QFA51, take a right turn on VZ, U1 for you Bay C23.
P4: Roger, right onto VZ, U1 for C23, QFA51.
C: BAW430, continue taxi, hold short of NC3.
P2: Hold short of NC3, BAW430.
P3: DLH15, request taxi.
C: DLH15, standby.
P5: Ground, SIA557, good evening.
C: SIA557, Ground, good evening, turn left following the greens to Bay D30.
P5: Follow the greens, D30, SIA557.
C: DLH15, make a right turn onto VY and hold short of WP.
P3: Right onto VY, hold short of WP, DLH15.
C: DLH15, continue make a right turn onto WP, follow the greens for holding position, Runway20L.

Questions
1. How many aircraft are involved in this long exchange?
2. Which of the airplanes in the exchange are inbound, and which are outbound?
3. What is controller's taxi instruction to QFA51?
4. What is the controller's taxi instruction to DLH15?

III. Look at the picture below and describe the procedures in detail.

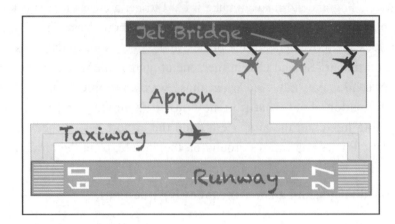

Part 2 Reading

Text A Intensive Reading

Air Traffic Control

I. INTRODUCTION

The increasing number of Aircrafts **proliferation** and increased use of air cause to use each country of the world Air Traffic Control radars to ensure the safety of passengers and Aircraft movement in the air. Every year advanced air traffic control equipment is introduced to protect the air and protect the safety of using Aircrafts. Air traffic control is also involved military action offence and defense in some countries. The air traffic control, its types and so of its advantage will be discussed here.

II. DEFINITIONS

1. Air Traffic Control

Air traffic control is a service provided by ground-based controllers who direct aircraft on the ground and through controlled **airspace**, and can provide **advisory** services to aircraft in non-controlled airspace. The primary purpose of Air Traffic Control worldwide is to prevent collisions, organize and **expedite** the flow of traffic, and provide information and other support for pilots. In some countries, Air Traffic Control plays a security or defensive role, or is operated by the military.

To prevent collisions, Air Traffic Control enforces **traffic separation rule**s which ensure each aircraft maintains a minimum amount of empty space around it at all times. Many aircraft also have collision avoidance systems which provide additional safety by warning pilots when other aircrafts get too close.

In many countries, Air Traffic Control provides services to all private, military, and commercial aircraft operating within its airspace. Depending on the type of flight and the class of airspace, Air Traffic Control may issue instructions that pilots are required to obey, or advisories (known as flight information in some countries) that pilots may be **at their discretion**, or **disregard**. The pilot in command is the final authority for the safe operation of the aircraft and may in an emergency, deviate from Air Traffic Control instructions to the extent required to maintain safe operation of their aircraft.

2. Aircraft

An aircraft is a machine that is able to fly by gaining support from the air, or, in general, the atmosphere of a planet.

3. Airspace

Airspace is the portion of the atmosphere controlled by a country above its territory, including its territorial waters, more generally, any specific three-dimensional portion of the atmosphere. It is not the same as aerospace, which is the general term for Earth's atmosphere and the outer space in its vicinity.

4. Air traffic Controllers

Air traffic controllers are people trained to maintain the safe, orderly and **expeditious** flow of air traffic in the global air traffic control system. The position of air traffic controller is one that requires highly specialized knowledge, skills, and abilities. Controllers apply separation rules to keep aircrafts at a safe distance from each other in their area of responsibility and move all aircraft safely and efficiently through their assigned **sector of airspace**, as well as on the ground. Because controllers have an incredibly large responsibility while on duty and make countless **split-second** decisions on a daily basis, the Air Traffic Control profession is consistently regarded around the world as one of the most mentally challenging careers, and can be **notoriously** stressful depending on many variables (equipment, configurations, weather, traffic volume, human factors, etc.).

5. Area Control Centre

Air Traffic Control provides services to aircraft in flight between airports as well. Pilots fly under one of two sets of rules for separation: Visual Flight Rules (VFR) or **Instrument Flight Rules (IFR)**. Air traffic controllers have different responsibilities to operating aircraft under the different sets of rules.

6. **En route**

En route air traffic controllers work in facilities called Air Traffic Control Centres, each of which is commonly referred to as a "Centre". Each centre is responsible for many thousands of square miles of airspace (known as a **Flight Information Region**) and for the airports within that airspace. Centres control IFR aircraft from the time they depart from an airport or terminal area's airspace to the time they arrive at another airport or terminal area's airspace. Centres may also "pick up" VFR aircraft that are already **airborne** and integrate them into the IFR system. These aircraft must, however, remain VFR until the Centre provides a **clearance**.

7. Radar coverage

Since centres control a large airspace area, they will typically use long range radar that has the capability, at higher altitudes, to see aircraft within 200 nautical miles (370 km) of the radar antenna. They may also use beacon radar data to control when it provides a better "picture" of the traffic or when it can fill in a portion of the area not covered by the long range radar.

III. TYPES OF AIR TRAFFIC CONTROL

1. Airport control

The primary method of controlling the immediate airport environment is visual observation from the airport control tower (TWR). The tower is a tall, windowed structure located on the airport grounds. Air traffic controllers are responsible for the separation and efficient movement of aircraft and vehicles operating on the taxiways and runways of the airport itself, and aircraft in the air near the airport, generally 5 to 10 nautical miles (9 to 18 km) depending on the airport **procedure**s.

Surveillance displays are also available to controllers at larger airports to assist with controlling air traffic. Controllers may use a radar system called **secondary surveillance radar** for airborne traffic approaching and **depart**ing. These displays include a map of the area, the position of various aircraft, and data **tag**s that include aircraft identification, speed, altitude, and other information described in local procedures. In adverse weather conditions the tower controllers may also use **Surface Movement Radar** (SMR), **Surface Movement Guidance and Control Systems** (SMGCS) or advanced SMGCS to control traffic on the **manoeuvring** area (taxiways and runway).

The areas of responsibility for TWR controllers fall into three general operational disciplines: Local Control or Air Control, Ground Control, and Flight Data/Clearance Delivery. Other categories, such as **Apron Control** or **Ground Movement Planner**, may exist at extremely busy airports.

While each TWR may have unique airport-specific procedures, such as multiple teams of controllers ('crews') at major or complex airports with multiple runways, the following provides a general concept of the **delegation** of responsibilities within the TWR environment. **Remote and Virtual Tower** (RVT) is a system based on Air Traffic Controllers being located somewhere other than at the local airport tower and still able to provide Air Traffic Control services. Displays for the Air Traffic Controllers may be either **optical live video** and/or **synthetic image**s based on surveillance sensor data.

2. Ground control

Ground Control (sometimes known as Ground Movement Control) is responsible for the airport "movement" areas, as well as areas not released to the airlines or other users. This generally includes all taxiways, inactive runways, holding areas, and some transitional aprons or **intersection**s where aircraft arrive, having **vacate**d the runway or departure gate. Exact areas and control responsibilities are clearly defined in local documents and agreements at each airport. Any aircraft, vehicle, or person walking or working in these areas is required to have clearance from Ground Control. This is normally done via **VHF/UHF radio**, but there may be special cases where other procedures are used. Aircraft or vehicles without radios must respond to ATC instructions via aviation light signals or else be led by vehicles with radios. People working on the airport surface normally have a communications link through which they can communicate with Ground Control, commonly either by handheld radio or even cell phone.

Ground Control is vital to the smooth operation of the airport, because this position impacts the sequencing of departure aircraft, affecting the safety and efficiency of the airport's operation. Some busier airports have Surface Movement Radar (SMR), such as, **ASDE-X** (Airport Surface Detection Equipment, is a runway-safety tool that enables air traffic controllers to detect potential runway conflicts by providing detailed coverage of movement on runways and taxiways. By collecting data from a variety of sources, ASDE-X is able to track vehicles and aircraft on airport surfaces and obtain identification information from aircraft transponders), **AMASS** (Airport Movement Area Safety System), designed to display aircraft and vehicles on the ground. These are used by Ground Control as an additional tool to control ground traffic, particularly at night or in poor visibility. There are a wide range of capabilities on these systems as they are being modernized. Older systems will display a map of the airport and the target. Newer systems include the capability to display higher quality mapping, radar target, data blocks, and safety alerts, and to interface with other systems such as **digital flight strip**s.

3. **Local control** or air control

Local Control (known to pilots as "Tower" or "Tower Control") is responsible for the active runway surfaces. Local Control **clear**s aircraft for takeoff or landing, ensuring that **prescribe**d runway separation will exist at all times. If Local Control detects any unsafe condition, a landing aircraft may be told to "go-around" and be re-sequenced into the **landing pattern** by the approach or terminal area controller. Within the TWR, a highly disciplined communications process between Local Control and Ground Control is an absolute necessity. Ground Control must request and gain approval from Local Control to cross any active runway with any aircraft or vehicle. Likewise, Local Control must ensure that Ground Control is aware of any operations that will impact the taxiways, and work with the approach radar controllers to create "holes" or "gaps" in the arrival traffic to allow taxiing traffic to cross runways and to allow departing aircraft to take off. **Crew Resource Management** (CRM) procedures are often used to ensure this communication process is efficient and clear, although this is not as prevalent as CRM for pilots.

4. Flight data / clearance delivery

Clearance Delivery is the position that issues route clearances to aircraft, typically before they **commence** taxiing. These contain details of the route that the aircraft is expected to fly after departure. Clearance Delivery or, at busy airports, the **Traffic Management Coordinator** (TMC) will, if necessary, coordinate with the en route centre and national command centre or flow control to obtain releases for aircraft. Often, however, such releases are given automatically or are controlled by local agreements allowing "**free-flow**" departures. When weather or extremely high demand for a certain airport or airspace becomes a factor, there may be ground "stops" (or "**slot** delays") or re-routes may be necessary to ensure the system does not get overloaded.

The primary responsibility of Clearance Delivery is to ensure that the aircrafts have the proper route and slot time. This information is also coordinated with the en route centre and Ground Control in order to ensure that the aircraft reaches the runway in time to meet the slot time provided by the command centre.

At some airports, Clearance Delivery also plans aircraft push-backs and engine starts, in which case it is known as the **Ground Movement Planner** (GMP): this position is particularly important at heavily congested airports to prevent taxiway and apron **gridlock**. Flight Data (which is routinely combined with Clearance Delivery) is the position that is responsible for ensuring that both controllers and pilots have the most current information: **pertinent** weather changes, **outage**s, airport ground delays/ground stops, runway closures, etc. Flight Data may inform the pilots using a recorded continuous **loop** on a specific frequency known as the **Automatic Terminal Information Service** (ATIS).

5. Approach and terminal control

Many airports have a radar control facility that is associated with the airport. In most countries, this is referred to as **terminal control**; in the U.S., it is referred to as a TRACON (**Terminal Radar Approach Control**). While every airport varies, terminal controllers usually handle traffic in a 30-to-50-nautical-mile (56 to 93 km) radius from the airport. Where there are many busy airports close together, one consolidated Terminal Control Centre may service all the airports. The airspace boundaries and altitudes assigned to a Terminal Control Centre, which vary widely from airport to airport, are based on factors such as traffic flows, neighbouring airports and terrain. A large and complex example is the London Terminal Control Centre which controls traffic for five main London airports up to 20,000 feet (6,100 m) and out to 100 nautical miles (190 km).

Terminal controllers are responsible for providing all ATC services within their airspace. Traffic flow is broadly divided into "departures", "arrivals", and "over flights". As aircraft move in and out of the terminal airspace, they are handed off to the next appropriate control facility (a control tower, an en-route control facility, or a **bordering** terminal or approach control). Terminal control is responsible for ensuring that aircraft are at an appropriate altitude when they are handed off, and that aircraft arrive at a suitable rate for landing.

Not all airports have a radar approach or terminal control available. In this case, the en-route centre or a neighbouring terminal or approach control may coordinate directly with the tower on the airport and **vector** inbound aircraft to a position from where they can land visually. At some of these airports, the tower may provide a non-radar procedural approach service to arriving aircraft handed over from a radar unit before they are visual to land. Some units also have a dedicated approach unit which can provide the procedural approach service either all the time or for any periods of radar outage for any reason. In the U.S., TRACONs are additionally designated by a three-letter **alphanumeric code**. For example, the Chicago TRACON is designated C90.

IV. CONCLUSION

Air Traffic Control is the system that directs the aircraft to be safe and makes rules to ensure the control of airspace and the safety of the aircrafts and their passengers. So every country in this world has controllers that controls its airspace. Even though some countries use Air Traffic Controllers to control their airspace as military and civilians, the most benefits of the Air Traffic Controllers are the safety. And although it is kind of developing, it is necessary to advance day by day.

New Words & Expressions

proliferation *n.* /prəˌlɪfəˈreɪʃn/	the sudden increase in the number or amount of sth; a large number of a particular thing 激增，涌现，增殖，大量的事物	
airspace *n.* /ˈeəspeɪs/	the part of the sky where planes fly, usually the part above a particular country that is legally controlled by that country 领空，（某国的）空域	
advisory *adj.* /ədˈvaɪzəri/	having the role of giving professional advice 顾问的，咨询的	
n.	an official warning that sth bad is going to happen 警报	
expedite *v.* /ˈekspədaɪt/	to make a process happen more quickly 加快，加速	
discretion *n.* /dɪˈskreʃn/	the freedom or power to decide what should be done in a particular situation 自行决定权	
	at (one's) discretion 自行决定	
disregard *v.* /ˌdɪsrɪˈgɑːd/	to not consider sth; to treat sth as unimportant 不理会，不顾，漠视	
expeditious *adj.* /ˌekspəˈdɪʃəs/	that works well without wasting time, money, etc. 迅速而有效的，迅速完成的	
sector *n.* /ˈsektə(r)/	a plane figure bounded by two radii	

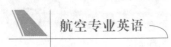

and the included arc of a circle; an area
or portion that is distinct from others,
especially that is under military control

扇形，部分，区域，地带（尤指军事管制的）

split-second *adj.* /ˌsplɪt ˈsekənd/　done very quickly or very accurately

一瞬间作出的，做得非常精确的

notoriously *adv.* /nəʊˈtɔːriəsli/　as is known to all

众所周知地

facility *n.* /fəˈsɪləti/　a place, usually including buildings, used
for a particular purpose or activity

（供特定用途的）场所

airborne *adj.* /ˈeəbɔːn/　(of a plane or passengers) in the air

升空

clearance *n.* /ˈklɪərəns/　official permission for a person or
vehicle to enter or leave an airport or a
country

（人、交通工具出入空港或出入境的）许可，
准许，飞行许可，放行许可

procedure *n.* /prəˈsiːdʒə(r)/　a way of doing sth, especially the usual
or correct way

程序，手续，步骤

surveillance *n.* /sɜːˈveɪləns/　careful watching of someone, especially
by an organization

监视，监控

depart *v.* /dɪˈpɑːt/　to leave a place, especially to start a trip

离场，离场起飞

tag *n.* /tæg/　a set of letters or symbols that are put
before and after a piece of text or data in
order to identify it or show that it is to be
treated in a particular way

标记，标签，标识符

manoeuvring *n.* /məˈnuːvərɪŋ/　a deliberate coordinated movement
requiring dexterity and skill

移动，部署

delegation *n.* /ˌdelɪˈgeɪʃn/　the act of empowering to act for another

委任，分配

synthetic *adj.* /sɪnˈθetɪk/　artificial; made by combining chemical
substances rather than being produced
naturally by plants or animals

人造的，（人工）合成的

intersection *n.* /ˌɪntəˈsekʃn/ a place where two or more roads, lines, etc. meet or cross each other
十字路口，交叉路口，交点

vacate *v.* /ˈveɪkeɪt/ to leave a building, seat, place, etc., especially so that sb / sth else can use it
搬出，腾出，空出（建筑物、座位等）

clear *v.* /klɪə(r)/ give official permission
批准

prescribed *adj.* /prɪˈskraɪbd/ conforming to set usage, procedure, or discipline
批准的，指定的

commence *v.* /kəˈmens/ to begin to happen; to begin sth
开始发生，开始，着手

slot *n.* /slɒt/ a position, a time or an opportunity for sb/sth, for example in a list, a programme of events or a series of broadcasts
位置，时间段，机会，席位，安排给航空器进离场的特定时间间隔

gridlock *n.* /ˈgrɪdlɒk/ a situation in which the traffic cannot move at all
交通堵塞

pertinent *adj.* /ˈpɜːtɪnənt/ relevant and/or appropriate to a particular situation
有关的，恰当的，相宜的

outage *n.* /ˈaʊtɪdʒ/ a period of time when the supply of electricity, etc. is not working
停电（等）期间

loop *n.* /luːp/ a set of instructions that is repeated again and again until a particular condition is satisfied
循环，回路，一套重复的指令，环路通信

border *v.* /ˈbɔːdə(r)/ to come very close to being sth.
靠近，接近，濒于，近乎

vector *n.* /ˈvektə(r)/ a course taken by an aircraft
（航空器的）航线

alphanumeric *adj.* /ˌælfənjuːˈmerɪk/
(also alphanumerical) containing both letters and numbers
有字母和数字的，字母与数字并用的

Air Traffic Control 空中交通管制

traffic separation rule	航路分离原则，航路分隔原则
sector of airspace	扇区
Instrument Flight Rules (IFR)	仪表飞行规则
en route	在途中，在路上
Flight Information Region	飞行情报区，飞航情报区
secondary surveillance radar	二次监视雷达，次级搜索雷达
Surface Movement Radar (SMR)	
	地面监察雷达，地面活动雷达
Surface Movement Guidance and Control Systems(SMGCS)	
	地面运动制导与控制系统
Apron Control / Ground Movement Planner	
	停机坪管制
Remote and Virtual Tower	远程虚拟塔台
optical live video	可视直播
synthetic image	合成图像
surveillance sensor data	监控传感数据
VHF (Very High Frequency) radio	
	手持对讲机，甚高频电台，超短波电台
UHF (Ultra High Frequency) radio	
	超短波无线电设备，超高频无线电设备，超短波无线电台
handheld radio	手持对讲机，无线电对讲机
Airport Surface Detection Equipment (ASDW-X)	
	机场地面雷达，机场地面探测设备
Airport Movement Area Safety System (AMASS)	
	机场活动区安全系统
digital flight strip	电子飞行进程单，电子飞行航线单
local control	机旁控制，现场控制
landing pattern	落地航线，着陆航线，着地前起落航线的飞行速率
Crew Resource Management	机员 / 机组资源管理
Traffic Management Coordinator	
	交通管理协调员
Automatic Terminal Information Service (ATIS)	
	自动终端信息业务
terminal control	终端管制，机场控制
Terminal Radar Approach Control (TRACON)	
	终端雷达进场控制
alphanumeric code	字母数字代码

Exercises

I. Answer the following questions.

1. What is Air Traffic Control?
2. What is the difference between airspace and aerospace?
3. Why are Air Traffic Controllers so challenging and stressful?
4. What is ground movement control?
5. What is the primary responsibility of Clearance Delivery?
6. What are terminal controllers mainly responsible for?

II. Multiple choice (one or more answers).

1. What is not the role of Air Traffic Control?
   A. To prevent collisions, organize and expedite the flow of traffic.
   B. To provide information and other support for pilots.
   C. To execute military security or defense.
   D. To perform search and rescue.
2. Which radar does controller use to guide airplane approaching and departing?
   A. Secondary surveillance radar.
   B. Surface Movement Radar.
   C. Long range radar.
   D. Beacon radar.
3. Which of the following is not the main operational disciplines that TWR controllers should follow?
   A. Local Control or Air Control.
   B. Ground Control.
   C. Flight Data/Clearance Delivery.
   D. Apron Control or Ground Movement Planner.
4. What kinds of surface movement radar do some busy airport use to operate the ground control?
   A. ASDE-X.
   B. AMASS.
   C. SMR.
   D. TMC.
5. Through which control, taxiway and apron gridlock could be prevented?
   A. GMP.
   B. ATIS.
   C. TMC.
   D. TRACON.
6. What is the normal traffic control radius from the airport of a terminal controller being responsible for?

A. 5-10 nautical mile.

B. 30-50 nautical mile.

C. 100 nautical mile.

D. 200 nautical mile.

Text B Extensive Reading

Airport Operation

Airports with an Operating Control Tower

1. When operating at an airport where traffic control is being exercised by a control tower, pilots are required to maintain two-way radio contact with the tower while operating within the Class B, Class C, and Class D surface area unless the tower authorizes otherwise. Initial callup should be made about 15 miles from the airport. Unless there is a good reason to leave the tower frequency before exiting the Class B, Class C, and Class D surface areas, it is a good operating practice to remain on the tower frequency for the purpose of receiving traffic information. In the interest of reducing tower frequency **congestion**, pilots are reminded that it is not necessary to request permission to leave the tower frequency once outside of Class B, Class C, and Class D surface areas. Not all airports with an operating control tower will have Class D airspace. These airports do not have weather reporting which is a requirement for surface based controlled airspace, previously known as a **control zone**. The controlled airspace over these airports will normally begin at 700 feet or 1,200 feet above ground level and can be determined from the **visual aeronautical chart**s. Pilots are expected to use good operating practices and communicate with the control tower as described in this section.

congestion *n.* /kənˈdʒestʃən/（交通）拥塞；塞车

control zone *n.* 航空控制区

visual aeronautical chart 目航视图

2. When necessary, the tower controller will issue clearances or other information for aircraft to generally follow the desired flight path (**traffic pattern**s) when flying in Class B, Class C, and Class D surface areas and the proper taxi routes when operating on the ground. If not otherwise authorized or directed by the tower, pilots of fixed-wing aircraft approaching to land

traffic pattern *n.* 起落航线

must circle the airport to the left. Pilots approaching to land in a helicopter must avoid the flow of fixed-wing traffic. However, in all instances, an appropriate clearance must be received from the tower before landing.

NOTE: This diagram is intended only to illustrate terminology used in identifying various components of a traffic pattern. It should not be used as a reference or guide on how to enter a traffic pattern. (See Figure 7.1)

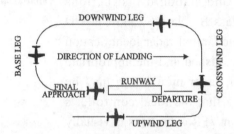

Figure 7.1 Components of a Traffic Pattern

3. The following terminology for the various components of a traffic pattern has been adopted as standard for use by control towers and pilots.

   A. **Upwind leg**. A flight path parallel to the landing runway in the direction of landing.

   B. **Crosswind leg**. A flight path at right angles to the landing runway off its takeoff end.

   C. **Downwind leg**. A flight path parallel to the landing runway in the opposite direction of landing.

   D. Base leg. A flight path at right angles to the landing runway off its approach end and extending from the downwind leg to the intersection of the extended runway centerline.

   E. **Final approach**. A flight path in the direction of landing along the extended runway centerline from the base leg to the runway.

   F. **Departure**. The flight path which begins after takeoff and continues straight ahead along the extended runway centerline. The departure climb continues until reaching a point at least 1/2 mile beyond the departure end of the runway and within 300 feet of the traffic pattern altitude.

4. Many towers are equipped with a tower **radar display**. The radar uses are intended to enhance the effectiveness and efficiency of the local control, or tower, position. They are not

leg *n.* /leg/ 航段

upwind *adv.* /ˌʌpˈwɪnd/ 逆风，顶风

upwind leg 起落航线第一边

crosswind *n.* /ˈkrɒswɪnd/ 侧风

crosswind leg 起落航线第二边

downwind *adv.* /ˌdaʊnˈwɪnd/ 顺风地，在下风处

downwind leg 起落航线第三边

base leg 起落航线第四边

final approach [航] 最终进场，降落路线的最后一个导航点

radar display 雷达显示器

intended to provide radar services or benefits to pilots except as they may **accrue** through a more efficient tower operation. The four basic uses are:

A. To determine an aircraft's exact location. This is accomplished by radar identifying the VFR aircraft through any of the techniques available to a radar position, such as having the aircraft squawk **ident**. Once identified, the aircraft's position and **spatial** relationship to other aircraft can be quickly determined, and standard instructions regarding VFR operation in Class B, Class C, and Class D surface areas will be issued. Once initial radar identification of a VFR aircraft has been established and the appropriate instructions have been issued, radar monitoring may be discontinued; the reason being that the local controller's primary means of surveillance in VFR conditions is visually scanning the airport and local area.

B. To provide radar traffic advisories. Radar traffic advisories may be provided to the extent that the local controller is able to monitor the radar display. Local control has primary control responsibilities to the aircraft operating on the runways, which will normally **supersede** radar monitoring duties.

C. To provide a direction or suggested **heading**. The local controller may provide pilots flying VFR with generalized instructions which will facilitate operations; e.g., "PROCEED SOUTHWESTBOUND, ENTER A RIGHT DOWNWIND RUNWAY THREE ZERO," or provide a suggested heading to establish radar identification or as an advisory aid to navigation; e.g., "SUGGESTED HEADING TWO TWO ZERO, FOR RADAR IDENTIFICATION." In both cases, the instructions are advisory aids to the pilot flying VFR and are not radar vectors.

NOTE: Pilots have complete discretion regarding acceptance of the suggested headings or directions and have sole responsibility for seeing and avoiding other aircraft.

D. To provide information and instructions to aircraft operating within Class B, Class C, and Class D surface areas. In an example of this situation, the local controller would use the radar to advise a pilot on an extended downwind when to turn base leg.

accrue *v.* /əˈkruː/ 增长，增加

ident *n.* /ˈaɪdent/ 识别，标识符，识别信号

spatial *adj.* /ˈspeɪʃl/ 空间的

supersede *v.* /ˌsjuːpəˈsiːd/ 取代，替代（已非最佳选择或已过时的事物）

heading *n.* /ˈhedɪŋ/ 前进方向

augment *v.* /ɔːgˈment/ 增加，提高，扩大

provision *n.* /prəˈvɪʒn/ 提供，供给，给养，供应品

NOTE: The above tower radar applications are intended to augment the standard functions of the local control position. There is no controller requirement to maintain constant radar identification. In fact, such a requirement could compromise the local controller's ability to visually scan the airport and local area to meet FAA responsibilities to the aircraft operating on the runways and within the Class B, Class C, and Class D surface areas. Normally, pilots will not be advised of being in radar contact since that continued status cannot be guaranteed and since the purpose of the radar identification is not to establish a link for the provision of radar services.

E. A few of the radar equipped towers are authorized to use the radar to ensure separation between aircraft in specific situations, while still others may function as limited radar approach controls. The various radar uses are strictly a function of FAA operational need. The facilities may be indistinguishable to pilots since they are all referred to as tower and no publication lists the degree of radar use. Therefore, when in communication with a tower controller who may have radar available, do not assume that constant radar monitoring and complete ATC radar services are being provided.

Traffic Patterns

1. It is recommended that aircraft enter the airport traffic pattern at one of the following altitudes listed below. These altitudes should be maintained unless another traffic pattern altitude is published in the Chart Supplement U.S. or unless otherwise required by the applicable distance from cloud criteria. (See Figure 7.2 and Figure 7.3):

A. Propeller-driven aircraft enter the traffic pattern at 1,000 feet above ground level (AGL).

B. Large and turbine-powered aircraft enter the traffic pattern at an altitude of not less than 1,500 feet AGL or 500 feet above the established pattern altitude.

C. Helicopters operating in the traffic pattern may fly a pattern similar to the fixed-wing aircraft pattern, but at a lower altitude (500 AGL) and closer to the runway. This pattern may be on the opposite side of the runway from fixed-wing traffic when airspeed requires or for practice power-off landings (**autorotation**) and if local policy permits. Landings not to the runway must avoid the flow of fixed wing traffic.

autorotation *n.*
/ˌɔːtəʊrəʊˈteɪʃən/ 自旋转

2. A pilot may vary the size of the traffic pattern depending on the aircraft's performance characteristics. Pilots of en route aircraft should be constantly alert for aircraft in traffic patterns and avoid these areas whenever possible.

3. Unless otherwise indicated, all turns in the traffic pattern must be made to the left, except for helicopters, as applicable.

4. On Sectional, Aeronautical, and VFR Terminal Area Charts, right traffic patterns are indicated at public-use and joint-use airports with the abbreviation "RP" (for Right Pattern), followed by the appropriate runway number(s) at the bottom of the airport data block.

EXAMPLE: RP 9, 18, 22R

*NOTE: Pilots are encouraged to use the standard traffic pattern. However, those pilots who choose to execute a **straight-in approach**, maneuvering for and execution of the approach should not disrupt the flow of arriving and departing traffic. Likewise, pilots operating in the traffic pattern should be alert at all times for aircraft executing straight-in approaches.*

straight-in approach 直线
进近

REFERENCE: AC 90-66B, Non-Towered Airport Flight Operations
RP indicates special conditions exist and refers pilots to the Chart Supplement U.S.
Right traffic patterns are not shown at airports with full-time control towers.

5. Wind conditions affect all airplanes in varying degrees. Figure 4 is an example of a chart used to determine the headwind, crosswind, and tailwind components based on wind direction and **velocity** relative to the runway. Pilots should refer to similar information provided by the aircraft manufacturer when determining these wind components.

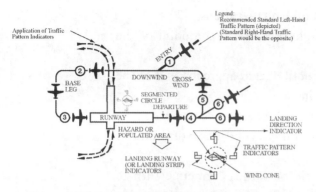

Figure 7.2 Traffic Pattern Operations Single Runway

EXAMPLE: Key to Traffic Pattern Operations

A. Enter pattern in level flight, **abeam** the midpoint of the runway, at **pattern altitude**.

B. Maintain pattern altitude until abeam approach end of the landing runway on downwind leg.

C. Complete turn to final at least 1/4 mile from the runway.

D. Continue straight ahead until beyond departure end of runway.

E. If remaining in the traffic pattern, commence turn to crosswind leg beyond the departure end of the runway within 300 feet of pattern altitude.

F. If departing the traffic pattern, continue straight out, or exit with a 45 degree turn (to the left when in a left-hand traffic pattern; to the right when in a right-hand traffic pattern) beyond the departure end of the runway, after reaching pattern altitude.

abeam *adj.* / *adv.* /əˈbiːm/ 横着的，指与船的龙骨或飞机机身成直角

pattern altitude *n.* 起落航线高度

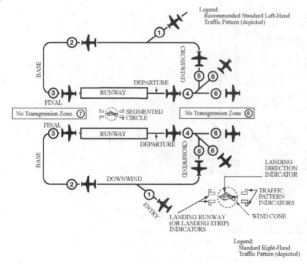

Figure 7.3 Traffic Pattern Operations Parallel Runway

EXAMPLE: Key to Traffic Pattern Operations

A. Enter pattern in level flight, abeam the midpoint of the runway, at pattern altitude.

B. Maintain pattern altitude until abeam approach end of the landing runway on downwind leg.

C. Complete turn to final at least 1/4 mile from the runway.

D. Continue straight ahead until beyond departure end of runway.

E. If remaining in the traffic pattern, commence turn to crosswind leg beyond the departure end of the runway within 300 feet of pattern altitude.

F. If departing the traffic pattern, continue straight out, or exit with a 45 degree turn (to the left when in a left-hand traffic pattern; to the right when in a right-hand traffic pattern) beyond the departure end of the runway, after reaching pattern altitude.

G. Do not **overshoot** final or continue on a track which will penetrate the final approach of the parallel runway.

overshoot *v.* /ˌəʊvəˈʃuːt/ 超过，越过（预定地点）

H. Do not continue on a track which will **penetrate** the departure path of the parallel runway.

penetrate *v.* /ˈpenɪtreɪt/ 穿过，进入

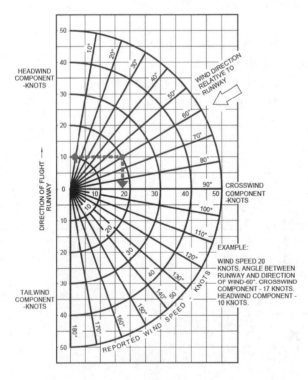

Figure 7.4 Headwind / Tailwind / Crosswind Component Calculator

Notes

1. **Airspace Classification of USA** 美国空域规范

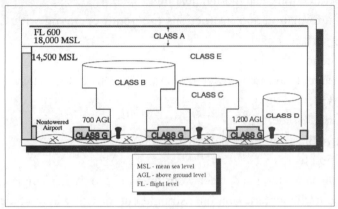

MSL—mean sea level

AGL—above ground level

FL—flight level

2. **Chart Supplement** 图表补充

Chart Supplements are a listing of data on record with the FAA on all open-to-the-public airports, seaplane bases, heliports, military facilities and selected private use airports specifically requested by the Department of Defense (DOD) for which a DOD instrument approach procedure has been published in the U.S. Terminal Procedures Publication, airport sketches, NAVAIDs, communications data, weather data sources, airspace, special notices, VFR waypoints, Airport Diagrams and operational procedures. Seven volumes cover the conterminous United States, Puerto Rico, and the Virgin Islands. The supplements include data that cannot be readily depicted in graphic form; e.g., airport hours of operation, types of fuel available, runway data, lighting codes, etc. The supplements are designed to be used in conjunction with charts and is published every 56 days. Volumes are side-bound 5 3/8 ×8 1/4 inches.

Exercises

I. Answer the following questions.

1. What is base leg?

2. Why are many towers equipped with radar?

3. How can an airport that doesn't have Class D airspace control aircraft?

4. Once receiving the instructions of suggested headings or directions, does pilot have any discretion on acceptance?

II. Multiple choice (one or more answers).

1. Which of the following is not the basic uses of radar?

A. To determine an aircraft's exact location.

B. To provide radar traffic advisories.

C. To provide a direction or suggested heading.

D. To provide information and instructions to aircraft operating with Class B, Class C, and Class D surface areas.

E. To issue clearances for aircraft.

2. What altitudes should a turbine-powered aircraft enter the traffic pattern?

A. 1,000 feet AGL.

B. Not less than 1,500 feet AEL.

C. 500 feet above the established pattern altitude.

D. Lower than 500 feet.

3. Put the following steps into the correct order for a pilot to fly in a single runway airport.

① If departing the traffic pattern, continue straight out, or exit with a 45 degree turn (to the left when in a left-hand traffic pattern; to the right when in a right-hand traffic pattern) beyond the departure end of the runway, after reaching pattern altitude.

② Enter pattern in level flight, abeam the midpoint of the runway, at pattern altitude.

③ Continue straight ahead until beyond departure end of runway.

④ If remaining in the traffic pattern, commence turn to crosswind leg beyond the departure end of the runway within 300 feet of pattern altitude.

⑤ Maintain pattern altitude until abeam approach end of the landing runway on downwind leg.

⑥ Complete turn to final at least 1/4 mile from the runway.

A. ③⑤④①⑥②

B. ①⑤③⑥④⑤

C. ②⑤⑥③④

D. ⑥③①⑤④②

4. In a parallel runway airport, if a pilot wants to depart the traffic pattern, what would he / she do?

A. Fly straight out.

B. Make a 45° left turn when in a right-hand traffic pattern.

C. Make a 45° right turn when in a left-hand traffic pattern.

D. Make a 90° left turn when in a left-hand traffic pattern.

Part 3 Translation

1. The Air Traffic Control (ATC) system is primarily to prevent a collision between aircrafts operating in the system and to organize and expedite the flow of traffic, it also has the capability to provide (with certain limitations) additional services, categorized as area control service, approach control service and aerodrome control service.

2. Air Traffic Flow Management (ATFM) is the regulation of air traffic in handling traffic in order to avoid exceeding airport or air traffic control capacity, and to ensure the effective use of available capacity.

3. Airspace shall be managed in the interest of the national security and with due consideration to the needs of civil and military aviation and the interests of the public and shall be planned in a unified manner to allow its rational, sufficient and effective utilization.

4. When an aircraft crosses a sector boundary, the responsibility for separating that aircraft passes on to the controller in the new sector. the original controller is known as the transferring controller, whereas the next controller is called the receiving controller. This transfer of separation responsibility is known as the transfer of control.

5. 机场是一个运输中心，用于飞机的降落和起飞，它不仅运输乘客，也为货物，如邮件，易腐货物和其他重要的物品提供运输。

6. 停机坪通常是铺有路面的，位于候机楼前或机库附近，是用作飞机停泊和装卸的地方。

7. 跑道是飞机在其上进行降落和起飞操作的平坦的道路，上面没有任何障碍物。

8. 许多机场都有灯光系统，帮助飞机在夜间、雨中或雾中使用跑道和滑行道。在跑道上，绿灯指示降落的起始位置，红灯指示跑道的结束位置。在滑行道上，蓝灯表示滑行道的边缘，一些机场在滑行道上也安装了绿灯，表示滑行道的中心线。

Part 4 Supplementary Reading

What Does An Air Traffic Controller Do?

Air Traffic Controllers are a critical part of commercial flight operations and safety. They can usually be found in a high tower near the airport, giving them a clear view of everything happening on the ground and in the skies above. But what exactly do Air Traffic Controllers do? Let's find out.

Keeping an Eye Out

Air Traffic Controllers (ATCs) are popularly known as the group who sit in a high tower over the airport, wearing headsets, and keeping an eye out over the airfield. This is an abbreviated definition of the job.

Indeed, any ATC's primary job is to monitor aircraft in its airspace and maintain constant communications with crew and ground staff. There are three types of main air traffic controllers, namely:

· Tower Controllers

· Approach and Departure Controllers

· En Route Controllers

Tower Controllers are responsible for the movement of planes, vehicles, and staff on runways and taxiways. They also receive and clear flight plans in advance, ensuring there is no overlap with the dozens of other flights on the same day. Moreover, they also communicate weather changes, runway closures, and more.

Once a plane is ready for departure, tower controllers will provide instructions and ensure no other planes or vehicles are obstructing the taxiways or exits. They also provide takeoff clearance, allowing a plane to safely depart without any complication. Overall, tower controllers manage traffic three miles to 30 miles in and around the airport.

However, it's not only planes that controllers have to handle. Vehicles such as safety cars, maintenance, and others must allow request permission before entering any active runway or taxiway.

In the Sky

Once a plane has safely taxied and taken off, approach and departure controllers take over communications. They are responsible for ensuring all planes maintain minimum separation while flying, that their flight path is clear of obstacles (including other aircraft), and handing off to en route controllers.

These are the controllers who can be seen behind radar screens, monitoring dozens of planes that might be in their airspace. In the US, these controllers work out of buildings known as Terminal Radar Approach Control Centers (TRACONs).

There are hundreds of TRACONs in total (147 to more precise), monitoring 50 kilometers of airspace each as designated by the FAA. Many of these are merged with existing airport ATCs, but some might be individual towers as well.

Any changes in altitude, speed, or direction must be cleared by controllers before being undertaken by pilots. The term "roger" is commonly used while confirming instructions over radio communications with controllers. Approach and departure controllers monitor planes up to 17,000 feet in altitude and about 20 to 50 miles from the airport.

As the name suggests, similar steps are repeated while the plane begins its approach into an airport. These controllers will once again check for separation, traffic, and flight plans before allowing the plane to land and handing back over to traffic controllers.

Cruising

Everything feels a bit more relaxed once the aircraft reaches cruising altitude. Passengers can recline their seats, cabin crew begin preparing for service, and planes are on autopilot. However, there are still air traffic controllers monitoring the plane, known as En Route controllers.

These controllers monitor planes as they fly over multiple states and follow their flight plans to their destination. In the US, they work out of Air Route Traffic Control Centers (ARTCCs), and the country is divided into 21 zones. Additionally, there are four combined control facilities (CCF) for oceanic and military traffic.

These ARTCCs monitor thousands of planes every day as they cross boundaries to reach their destinations. A transcontinental flight can easily pass through several ARTCCs on its way. Their primary job is to guide planes along their flight path, prevent any loss of separation and collisions, and general safety of the airspace. Planes spend most of their time in contact ARTCCs as they travel hundreds or thousands of miles every flight.

There are also nation-level airspace controllers who monitor traffic for any large-scale bottlenecks. There is only one in the US, the Air Traffic Control Systems Command Center (ATCSCC), and they don't communicate directly with pilots. Instead, this facility focuses on communicating with other ATC towers to ensure a smooth flow of traffic or warn of security concerns.

Important Job

Air Traffic Controllers play a crucial role in aviation and work to prevent any kind of safety incidents. From ensuring two wingtips don't graze on the ground to preventing any mid-air collisions, controllers have a heavy responsibility from their perch in the sky. Considering airports can see hundreds or thousands of flights daily, vigilance is crucial.

For this reason, air traffic controllers have strict entry requirements. In the US, you can only become a controller before your 31st birthday and have to retire by 56. This is to ensure sharp mental acuity, situational awareness, and short-term memory. Moreover, any serious health conditions like heart disease, diabetes, or mental illness can be disqualifying.

Overall, the air traffic controllers must be quick to respond to a crisis and are held to similar health standards as pilots. This is because even the smallest of delays or hesitation can result in a disaster during a crisis. This means being an air traffic controller can be a stressful job at times, especially when traffic is high. However, it is undoubtedly rewarding and a critical job for air travel to continue smoothly every day.

Unit 8　Aviation Weather

Lead-in

　　　A weather brief should be part of any preparation for flight. Here it is important to know the way around aviation weather reporting – what is available and how is it read.　In USA, in aviation, weather service is a combined effort of the National Weather Service (NWS), the Federal Aviation Administration (FAA), the Department of Defense (DOD), and other aviation groups and individuals. Because of the increasing need for worldwide weather services, foreign weather organizations also provide vital input. While weather forecasts are not 100 percent accurate, meteorologists, through careful scientific study and computer modeling, have the ability to predict the weather patterns, trends, and characteristics with increasing accuracy. Through a complex system of weather services, government agencies, and independent weather observers, pilots and other aviation professionals receive the benefit of this vast knowledge base in the form of up-to-date weather reports and forecasts. These reports and forecasts enable pilots to make informed decisions regarding weather and flight safety.

　　This unit will cover:
- Aviation weather reports
- Aviation weather abbreviations
- The impact of weather on aviation

航空专业英语

Part 1 Listening and Speaking

New Words & Expressions

atmosphere *n.* /ˈætməsfɪə(r)/ — the mixture of gases that surrounds the earth
大气，大气层，大气圈

mass *n.* /mæs/ — a large amount of a substance that does not have a definite shape or form
团，块，堆

definite *adj.* /ˈdefɪnət/ — easily or clearly seen or understood; obvious
清楚的，明显的

apparent *adj.* /əˈpærənt/ — easy to see or understand; obvious
显而易见，明白易懂，显然

groundspeed *n.* /ˈɡraʊndspiːd/ — the speed of an aircraft relative to the ground
对地速率

progress *n.* /ˈprəʊɡres/ — movement forwards or towards a place
前进，行进

path *n.* /pɑːθ/ — a line along which sb/sth moves; the space in front of sb/sth as they move
路线，道路，航径

track *n.* /træk/ — the path or direction that sb/sth is moving in
路径，路线，方向，飞行航线，航线

course *n.* /kɔːs/ — a direction or route followed by a ship or an aircraft
航向，航线

cumulonimbus *n.* /ˌkjuːmələʊˈnɪmbəs/ — a high mass of thick cloud with a flat base, often seen during thunderstorms
积雨云（常伴有雷阵雨）

trough *n.* /trɒf/ — a long narrow region of low air pressure between two regions of higher pressure
低压槽，槽形低压

precipitation *n.* /prɪˌsɪpɪˈteɪʃn/ — rain, snow, etc. that falls; the amount of this that falls
降水，降水量（包括雨、雪、冰等）

precipitation area — 降水区

140

ceiling	最高飞行限度，升限
trough line	[气象] 槽线，槽底线
squall line	[气象] 飑线，阵风线
sequence report	定时天气报告

I. Filling the blanks with the words from the box.

| ① 20 | ② 100 | ③ 120 | ④ 140 |
| ⑤ eastward | ⑥ south | ⑦ right | ⑧ combining |

Effect of Wind

As discussed in the study of the atmosphere, wind is a mass of air moving over the surface of the Earth in a definite direction. When the wind is blowing from the north at 25 knots, it simply means that air is moving southward over the Earth's surface at the rate of 25 NM (nautical miles) in 1 hour.

Under these conditions, any inert object free from contact with the Earth is carried 25 NM southward in 1 hour. This effect becomes apparent when such things as clouds, dust and toy balloons are observed being blown along by the wind. Obviously, an aircraft flying within the moving mass of air is similarly affected.

As shown in the figure, an aircraft flying _____ at an airspeed of 120 knots in still air has a groundspeed exactly the same – _____ knots. If the mass of air is moving eastward at _____ knots, the airspeed of the aircraft is not affected, but the progress of the aircraft over the ground is 120 plus 20, or a groundspeed of _____ knots. On the other hand, if the mass of air is moving westward at 20 knots, the airspeed of the aircraft remains the same, but groundspeed becomes 120 minus 20, or _____ knots.

Assuming no correction is made for wind effect, if an aircraft is heading eastward at 120 knots, and the air mass moving southward at 20 knots, the aircraft at the end of 1 hour is almost 120 miles east of its point of departure because of its progress through the air. It is 20 miles _____ because of the motion of the air. Under these circumstances, the airspeed remains 120 knots, but the groundspeed is determined by _____ the movement of the aircraft with that of the air mass. GS (groundspeed) can be measured as the distance from the point of departure to the position of the aircraft at the end of 1 hour.

The GS can be computed by the time required to fly between two points, a known distance apart. The direction in which the aircraft is pointing as it flies is heading. Its actual path over the ground, which is a combination of the motion of the aircraft and the motion of the air, is its track. The angle between the heading and the track is drift angle. If the aircraft heading coincides with the true course and the wind is blowing from the left, the track does not coincide with the true course. The wind causes the aircraft to drift to the _____, so the track falls to the right of the desired course or true course.

II. Listen to the dialogue and answer the questions following.

F – Weather Forecaster   P – Pilot

P: Good morning, sir. I am planning a flight to Beijing. Could you tell me the weather at Beijing?

F: Morning, captain, I am afraid your flight has to be delayed due to a severe thunderstorm over Beijing city.

P: I am sorry to hear that. Please tell me something more about it.

F: There are some cumulonimbus clouds building up hundreds miles south. Our radar shows some cells. A squall line might be developing ahead of the front.

P: Maybe there is heavy rain.

F: I agree with you. Look at the upper chart. There is a trough line across the route near Beijing city. Precipitation area is moving towards the south.

P: I am sure the area over there has low ceiling and poor visibility.

F: Yes, I think so. As a matter of fact, many of the airports in that area are below landing minimums due to poor visibility.

P: Is there any latest weather report, sir?

F: Sure, here is a sequence report for Beijing Airport. An hourly weather report is very helpful.

P: I know what you mean. I have to wait at least one hour.

F: I am afraid so.

Questions

1. Where and what are they talking about?
2. What does the upper chart show?
3. What are the words that are related to thunderstorm in the exchange?
4. What will the pilot do?

III. Look at the picture below and describe the effect of wind on plane while flying.

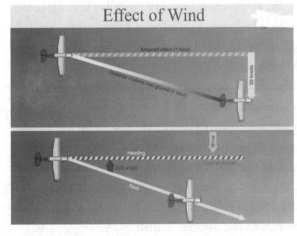

key words:
· airspeed effect (1 hour)
· 20 knots
· Distance covered over ground (1 hour)
· heading
· desired course
· drift angle
· track

Part 2 Reading

Text A Intensive Reading

Aviation Weather Reports

Aviation weather reports are designed to give accurate depictions of current weather conditions. Each report provides current information that is updated at different times. Some typical reports are aviation routine weather reports (METAR), pilot weather reports (PIREPs), and radar weather reports (RAREPs).

Aviation Routine Weather Report (METAR)

An Aviation Routine Weather Report, or METAR, is an observation of current **surface weather** reported in a standard international format. While the METAR **code** has been adopted worldwide, each country is allowed to make **modification**s to the code. Normally, these differences are minor but necessary to **accommodate** local procedures or particular units of measure.

METAR KGGG 161753Z AUTO 14021G26 3/4SM +TSRA BR BKN008 OVC012CB 18/17 A2970 RMK PRESFR

A typical METAR report contains the following information in **sequential** order:

1. Type of report – There are two types of METAR reports. The first is the routine METAR report that is transmitted every hour. The second is the **Aviation Selected Special Weather Report** (SPECI). This is a special report that can be given at any time to update the METAR for rapidly changing weather conditions, aircraft **mishap**s, or other critical information.

2. **Station identifier** – Each station is identified by a four-letter code as established by the **International Civil Aviation Organization** (ICAO). In the 48 **contiguous** states, a unique three-letter identifier is **preceded** by the letter "K". For example, Gregg County Airport in Longview, Texas, is identified by the letters "KGGG", K being the country **designation** and GGG being the airport identifier. In other regions of the world, including Alaska and Hawaii, the first two letters of the four-letter ICAO identifier indicate the region, country, or state. Alaska identifiers always begin with the letters "PA" and Hawaii identifiers always begin with the letters "PH". A list of station identifiers can be found at an **FSS** or **NWS** office.

3. Date and time of report – The date and time (161753Z) are **depict**ed in a six-digit group. The first two digits of the six-digit group are the date. The last four digits are the time of the METAR, which is always given in **Coordinated Universal Time** (UTC). A "Z" is **append**ed to the end of the time to denote the time is given in **Zulu Time** as opposed to local time.

4. **Modifier** – Modifiers **denote** that the METAR came from an **automated** source or that

the report was corrected. If the **notation** "AUTO" is listed in the METAR, the report came from an automated source. It also lists "AO1" or "AO2" in the **remark**s section to indicate the type of **precipitation** sensors employed at the automated station. When the modifier "COR" is used, it identifies a corrected report sent out to replace an earlier report that contained an error.Example: METAR KGGG 161753Z COR.

5. Wind – Winds are reported with five digits (14021) unless the speed is greater than 99 knots, in which case the wind is reported with six digits. The first three digits indicate the direction the wind is blowing in tens of degrees. If the wind is **variable**, it is reported as "VRB". The last two digits indicate the speed of the wind in knots (KT) unless the wind is greater than 99 knots, in which case it is indicated by three digits. If the winds are **gusting**, the letter "G" follows the windspeed (G26). After the letter "G", the **peak gust** recorded is provided. If the wind varies more than 60° and the windspeed is greater than 6 knots, a separate group of numbers, separated by a "V", will indicate the **extreme**s of the wind directions.

6. Visibility – The **prevailing** visibility (3/4 SM) is reported in **statute mile**s as denoted by the letters "SM". It is reported in both miles and **fraction**s of miles. At times, RVR, or **runway visual range** is reported following the **prevailing visibility**. RVR is the distance a pilot can see down the runway in a moving aircraft. When RVR is reported, it is shown with an R, then the runway number followed by a **slant**, then the visual range in feet. For example, when the RVR is reported as R17L/1400FT, it translates to a visual range of 1,400 feet on runway 17 left.

7. Weather – Weather can be broken down into two different categories: **qualifier**s and weather phenomenon (+TSRA BR). First, the qualifiers of **intensity**, **proximity**, and the **descriptor** of the weather will be given. The intensity may be light (-), moderate (blank), or heavy (+). Proximity only depicts weather phenomena that are in the airport **vicinity**. The notation "VC" indicates a specific weather phenomenon is in the vicinity of 5 to 10

Qualifier		Weather Phenomena		
Intensity or Proximity 1	Descriptor 2	Precipitation 3	Obscuration 4	Other 5
- Light	MI Shallow	DZ Drizzle	BR Mist	PO Dust/sand whirls
Moderate (no qualifier)	BC Patches	RA Rain	FG Fog	SQ Squalls
+ Heavy	DR Low Drifting	SN Snow	FU Smoke	FC Funnel cloud
VC in the vicinity	BL Blowing	SG Snow grains	DU Dust	+FC Tornado or Waterspout
	SH Showers	IC Ice Crystals (diamond dust)	SA Sand	SS Sandstorm
	TS Thunderstorms	PL Ice Pellets	HZ Haze	DS Dust storm
	FZ Freezing	GR Hail	PY Spray	
	PR Partial	GS Small hail or snow pellets	VA Volcanic ash	
		UP *Unknown Precipitation		

The weather groups are constructed by considering columns 1-5 in this table, in sequence; i.e., intensity, followed by descriptor, followed by weather phenomena; i.e., heavy rain showers(s) is coded as +SHRA.
* Automated stations only

Figure 8.1 Descriptors and Weather Phenomena Used in a TypicalMETAR.

miles from the airport. Descriptors are used to describe certain types of precipitation and **obscurations**. Weather phenomena may be reported as being precipitation, obscurations, and other phenomena such as **squalls** or **funnel cloud**s. Descriptions of weather phenomena as they begin or end, and **hailstone** size are also listed in the remarks sections of the report.

8. Sky condition – Sky condition (BKN008 OVC012CB) is always reported in the sequence of amount, height, and type or indefinite ceiling/height (vertical visibility). The heights of the **cloud base**s are reported with a three-digit number in hundreds of feet above the ground. Clouds above 12,000 feet are not detected or reported by an automated station. The types of clouds, specifically **towering cumulus** (TCU) or cumulonimbus (CB) clouds, are reported with their height. Contractions are used to describe the amount of cloud coverage and **obscuring** phenomena. The amount of sky coverage is reported in eighths of the sky from horizon to horizon.

Sky Cover	Less than 1/8 (Clear)	1/8-2/8 (Few)	3/8-4/8 (Scattered)	5/8-7/8 (Broken)	8/8 or Overcast (Overcast)
Contraction	SKC CLR FEW	FEW	SCT	BKN	OVC

Figure 8.2 Reportable Contractions for Sky Condition

9. Temperature and **dewpoint** – The air temperature and dewpoint are always given in degrees Celsius (18/17). Temperatures below 0°C are preceded by the letter "M" to indicate minus.

10. **Altimeter setting** – The altimeter setting is reported as inches of mercury in a four-digit number group (A2970). It is always preceded by the letter "A." Rising or falling pressure may also be denoted in the remarks sections as "**PRESRR**" or "**PRESFR**" respectively.

11. Remarks – Comments may or may not appear in this section of the METAR. The information contained in this section may include wind data, **variable visibility**, beginning and ending times of particular phenomenon, pressure information, and various other information **deem**ed necessary. An example of a remark regarding weather phenomenon that does not fit in any other category would be: OCNL LTGICCG. This translates as **occasional** lightning in the clouds, and from cloud to ground. Automated stations also use the remarks section to indicate the equipment needs maintenance. The remarks section always begins with the letters "RMK."

Example: METAR BTR 161753Z 14021G26 3/4SM -RA BR BKN008 OVC012 18/17 A2970 RMK PRESFR

Explanation:

Type of Report:	Routine METAR
Location:	Baton Rouge, Louisiana
Date:	16th day of the month

Time:	1753 Zulu
Modifier:	None shown
Wind Information:	Winds 140° at 21 knots gusting to 26 knots
Visibility:	3/4 statute mile
Weather:	light rain and mist
Sky Conditions:	Skies broken 800 feet, overcast 1,200
Temperature:	Temperature 18°C, dewpoint 17°C
Altimeter:	29.70 in. Hg.
Remarks:	Barometric pressure is falling.

Pilot Weather Reports (PIREPs)

Pilot weather reports provide valuable information regarding the conditions as they actually exist in the air, which cannot be gathered from any other source.

Pilots can confirm the height of bases and tops of clouds, locations of **wind shear** and **turbulence**, and the location of in-flight icing. If the ceiling is below 5,000 feet, or visibility is at or below 5 miles, ATC facilities are required to **solicit** PIREPs from pilots in the area.

When unexpected weather conditions are encountered, pilots are encouraged to make a report to an FSS or ATC. When a pilot weather report is filed, the ATC facility or FSS will add it to the **distribution** system to **brief** other pilots and provide in-flight **advisories**.

PIREPs are easy to file and a standard reporting form outlines the manner in which they should be filed.

Encoding Pilot Weather Reports (PIREPS)			
1	XXX	3-letter station identifier	Nearest weather reporting location to the reported phenomenon
2	UA	Routine PIREP, UUA-Urgent PIREP.	
3	/OV	Location	Use 3-letter NAVAID idents only. a. Fix: /OV ABC, /OV ABC 090025. b. Fix: /OV ABC 045020-DEF, /OV ABC-DEF-GHI
4	/TM	Time	4 digits in UTC: /TM 0915.
5	/FL	Altitude/flight level	3 digits for hundreds of feet. If not known, use UNKN: /FL095, /FL310, /FLUNKN.
6	/TP	Type aircraft	4 digits maximum. If not known, use UNKN: /TP L329, /TP B727, /TP UNKN.
7	/SK	Sky cover/cloud layers	Describe as follows: a. Height of cloud base in hundreds of feet. If unknown, use UNKN. b. Cloud cover symbol. c. Height of cloud tops in hundreds of feet.
8	/WX	Weather	Flight visibility reported first: Use standard weather symbols: /WX FV02SM RA HZ, /WX FV01SM TSRA.
9	/TA	Air temperature in celsius (C)	If below zero, prefix with a hyphen: /TA 15, /TA M06.
10	/WV	Wind	Direction in degrees magnetic north and speed in six digits: /WV270045KT, WV 280110KT.
11	/TB	Turbulence	Use standard contractions for intensity and type (use CAT or CHOP when appropriate). Include altitude only if different from /FL, /TB EXTRM, /TB LGT-MOD BLO 090.
12	/IC	Icing	Describe using standard intensity and type contractions. Include altitude only if different than /FL: /IC LGT-MOD RIME, /IC SEV CLR 028-045.
13	/RM	Remarks	Use free form to clarify the report and type hazardous elements first: /RM LLWS -15KT SFC-030 DURC RY22 JFK.

Figure 8.3 PIREP **Encoding** and **Deconding**

Figure 8.3 shows the elements of a PIREP form. Item numbers one through five are required information when making a report, as well as at least one weather phenomenon encountered. PIREPs are normally transmitted as an individual report, but may be appended to a surface report. Pilot reports are easily decoded and most contractions used in the reports are self-explanatory.

Example:UA/OV GGG 090025/ M 1450/ FL 060/ TP C182/ SK 080 OVC/ WX FV 04R/
TA 05/ WV 270030/ TB LGT/ RM HVY RAIN

Explanation:

Type:	Routine pilot report
Location:	25 NM out on the 090° radial, Gregg County VOR
Time:	1450 Zulu
Altitude or Flight Level:	6,000 feet
Aircraft Type:	Cessna 182
Sky Cover:	8,000 overcast
Visibility/Weather:	4 miles in rain
Temperature:	5° Celsius
Wind:	270° at 30 knots
Turbulence:	Light
Icing:	None reported
Remarks:	Rain is heavy.

Radar Weather Reports (RAREPs)

Areas of precipitation and thunderstorms are observed by radar **on a routine basis**. Radar weather reports are **issue**d by radar stations at 35 minutes past the hour, with special reports issued as needed.

Radar weather reports provide information on the type, intensity, and location of the **echo** top of the precipitation.

SYMBOL	MEANING
R	rain
RW	rain shower
S	snow
Sw	snow shower
T	thunderstorm
SYMBOL	**INTENSITY**
-	light
(none)	moderate
+	heavy
++	very heavy
X	intense
XX	extreme
CONTRACTION	**OPERATIONAL STATUS**
PPINE	Radar is operating normally, but there are no echoes being detected.
PPINA	Radar observation is not available.
PPIOM	Radar is inoperative or out of service.
AUTO	Automated radar report from WSR-88D.

Figure 8.4 Radar weather report codes

These reports may also include direction and speed of the area of precipitation as well as the height and base of the precipitation in hundreds of feet **MSL**. RAREPs are especially valuable for preflight planning to help avoid areas of severe weather. However, radar only detects objects in the atmosphere that are large enough to be considered precipitation. Cloud bases and **cloud top**s, ceilings, and visibility are not detected by radar.

A typical RAREP will include:

· Location identifier and time of radar observation.
· **Echo pattern**:
  1. Line (LN) – A line of precipitation echoes at least 30 miles long, at least four times as long as it is wide, and at least 25 percent coverage within the line.
  2. Area (AREA) – A group of echoes of similar type and not classified as a line.
  3. Single Cell (CELL) – A single isolated **convective** echo such as a rain shower.
· Area coverage in tenths.
· Type and intensity of weather.
· **Azimuth**, referenced to **true north**, and range, in **nautical mile**s, from the radar site, of points defining the echo pattern. For lines and areas, there will be two azimuth and range sets that define the pattern. For cells, there will be only one azimuth and **range set**.
· **Dimension** of echo pattern – The dimension of an echo pattern is given when the azimuth and range define only the center line of the pattern.
· Cell movement – Movement is only coded for cells; it will not be coded for lines or areas.
· Maximum top of precipitation and location. Maximum tops may be coded with the symbols "MT" or "MTS." If it is coded with "MTS," it means that satellite data as well as radar information was used to measure the top of the precipitation.
· If the word "AUTO" appears in the report, it means the report is automated from **WSR-88D weather radar** data.
· The last section is primarily used to prepare **radar summary chart**s, but can be used during preflight to determine the maximum precipitation intensity within a specific **grid box**. The higher the number, the greater the intensity. Two or more numbers appearing after a grid box reference, such as PM34, indicates precipitation in **consecutive** grid boxes.

Example: TLX 1935 LN 8 TRW++ 86/40 199/115 20W C2425 MTS 570 AT 159/65 AUTO ^MO1 NO2 ON3 PM34 QM3 RL2=

Explanation:

The radar report gives the following information: The report is automated from Oklahoma City and was made at 1935 UTC. The echo pattern for this radar report indicates a line of echos covering 8/10ths of the area.

Thunderstorms and very heavy rain showers are indicated. The next set of numbers indicate the azimuth that defines the echo (86° at 40 NM and 199° at 115 NM). Next, the dimension of this echo is given as 20 nautical miles wide (10 nautical miles on either side of the line defined by the azimuth and range). The cells within the line are moving from 240° at 25 knots. The maximum top of the precipitation, as determined by radar and satellite, is 57,000 feet and it is located on the 159° radial, 65 NM out. The last line indicates the intensity of the precipitation, for example in grid QM the intensity is 3 or heavy precipitation. (1 is light and 6 is extreme.)

code *n.* /kəʊd/ (often in compounds) a system of words, letters, numbers or symbols that represent a message or record information secretly or in a shorter form
密码，暗码，电码，代码

modification *n.* /ˌmɒdɪfɪˈkeɪʃn/ the act or process of changing sth in order to improve it or make it more acceptable; a change that is made
修改，改进，改变

accommodate *v.* /əˈkɒmədeɪt/ be agreeable or acceptable to
顺应，适应

sequential *adj.* /sɪˈkwenʃl/ following in order of time or place
按次序的，顺序的，序列的

mishap *n.* /ˈmɪshæp/ a small accident or piece of bad luck that does not have serious results
小事故，晦气

identifier *n.* /aɪˈdentɪfaɪə(r)/ a series of characters used to refer to a program or set of data within a program
标识符，标识号，识别字

contiguous *adj.* /kənˈtɪgjʊəs/ touching or next to sth
相接的，相邻的

precede *v.* /prɪˈsiːd/ to happen before sth or come before sth/sb in order
在……之前发生（或出现），先于

designation *n.* /ˌdezɪgˈneɪʃn/ the action of choosing a person or thing for a particular purpose, or of giving them or it a particular status
选定，指定，委任

depict *v.* /dɪˈpɪkt/ to describe sth in words, or give an impression of sth in words or with a picture
描写，描述，刻画

append *v.* /əˈpend/ to add sth to the end of a piece of writing
（在文章后面）附加，增补

modifier *n.* /ˈmɒdɪfaɪə(r)/ a word, such as an adjective or adverb, that describes another word or group of words, or restricts its/their meaning in some way
修饰语

denote *v.* /dɪˈnəʊt/ to mean sth
表示，意指

automated *adj.* /ˈɔːtəmeɪtɪd/ operated by automation
自动的

notation *n.* /nəʊˈteɪʃn/ a system of signs or symbols used to represent information, especially in mathematics, science and music
（数学、科学和音乐中的）符号，记号，谱号

remark *n.* /rɪˈmɑːk/ explicit notice
备注，评论

precipitation *n.* /prɪˌsɪpɪˈteɪʃn/ rain, snow, etc. that falls; the amount of this that falls
降水，降水量（包括雨、雪、冰等）

gust *v.* /ɡʌst/ (of the wind) to suddenly blow very hard
猛刮，劲吹

extreme *n.* /ɪkˈstriːm/ the greatest or highest degree of sth
极端，极度，极限

prevailing *adj.* /prɪˈveɪlɪŋ/ existing or most common at a particular time
普遍的，盛行的，流行的

fraction *n.* /ˈfrækʃn/ a division of a number, for example 5/8
分数，小数

slant *n.* /slɑːnt/ a crooked line
斜线，斜杠

qualifier *n.* /ˈkwɒlɪfaɪə(r)/ a word, especially an adjective or adverb, that describes another word in a particular way
修饰词（尤指形容词或副词）

intensity *n.* /ɪnˈtensəti/ the strength of sth, for example light, that can be measured
强度，烈度

proximity *n.* /prɒkˈsɪməti/ the state of being near sb/sth in distance or time

（时间或空间）接近，邻近，靠近

descriptor *n.* /dɪˈskrɪptə/
a word or expression used to describe or identify sth
描述符

vicinity *n.* /vəˈsɪnəti/
the area around a particular place
周围地区，邻近地区，附近

obscuration *n.* /ˌɒbskjʊˈreɪʃən/
dimness, gloom
昏暗

squall *n.* /skwɔːl/
a sudden strong and violent wind, often during rain or snow storms
飑（暴风雨或暴风雪中突起的狂风）

hailstone *n.* /ˈheɪlstəʊn/
a small ball of ice that falls like rain
雹块，雹子

obscuring *adj.* /əbˈskjʊərɪŋ/
not clear, invisible
模糊的

dewpoint *n.* /djʊˈpɔɪnt/
the temperature at which air can hold no more water, below this temperature the water comes out of the air in the form of drops.
露点（空气中水气含量达到饱和的气温，低于此温度时水气从空气中析出凝成水珠）

altimeter *n.* /ˈæltɪmiːtə(r)/
an instrument for showing height above sea level, used especially in an aircraft
（尤指用于飞行器中的）测高仪，高度表

deem *v.* /diːm/
(not usually used in the progressive tenses) to have a particular opinion about sth
（通常不用于进行时）认为，视为，相信

occasional *adj.* /əˈkeɪʒənl/
happening or done sometimes but not often
偶尔的，偶然的，临时的

barometric *adj.* /ˌbærəˈmetrɪk/
barometric pressure is the atmospheric pressure that is shown by a barometer
气压表的

turbulence *n.* /ˈtɜːbjələns/
a series of sudden and violent changes in the direction that air or water is moving in
（空气和水的）湍流，涡流，紊流

solicit *v.* /səˈlɪsɪt/
to ask sb for sth, such as support, money, or information; to try to get sth or

persuade sb to do sth

索求，请求……给予（援助、钱或信息），征求，筹集

distribution *n.* /ˌdɪstrɪˈbjuːʃn/　the act of giving or delivering sth to a number of people

分发，分送

brief *v.* /briːf/　to give sb information about sth so that they are prepared to deal with it

给（某人）指示，向（某人）介绍情况

advisory *n.* /ədˈvaɪzəri/　an official warning that sth bad is going to happen

警报

encode *v.* /ɪnˈkəʊd/　to change ordinary language into letters, symbols, etc. in order to send secret messages

把……译成电码（或密码）

decode *v.* /ˌdiːˈkəʊd/　to find the meaning of sth, especially sth. that has been written in code

解（码），破译（尤指密码）

contraction *n.* /kənˈtrækʃn/　the process of becoming smaller; a short form of a word

收缩，缩小，词的缩约形式

radial *n.* /ˈreɪdiəl/　径向，雷达定位

overcast *adj.* /ˌəʊvəˈkɑːst/　covered with clouds; dull

阴天的，多云的，阴暗的

issue *v.* /ˈɪʃuː/　to make sth known formally

宣布，公布，发出

echo *n.* /ˈekəʊ/　the reflecting of sound off a wall or inside a confined space so that a noise appears to be repeated; a sound that is reflected back in this way

回响，回声，回音

convective *adj.* /kənˈvektɪv/　the state that warmer parts move up and colder parts move down

对流的，传递性的

azimuth *n.* /ˈæzɪməθ/　the angular distance usually measured clockwise from the north point of the horizon to the intersection with the horizon of the vertical circle passing through a celestial body

方位角

dimension *n.* /daɪˈmenʃn/ the size and extent of a situation
规模，程度，范围

consecutive *adj.* /kənˈsekjətɪv/ following one after another in a series, without interruption
连续不断的

Aviation Routine Weather Report (METAR)
气象常规天气报告

surface weather 地面气象

Aviation Selected Special Weather Report (SPECI)
飞航特别天气报告

station identifier 航站标识符

International Civil Aviation Organization
国际民用航空组织，国际民航组织

Flight Service Station (FSS) 飞行服务站

National Weather Service (NWS)（美）国家气象局

Coordinated Universal Time(UTC) 协调世界时，协调通用时间，协调宇宙时，道界时

Zulu Time 祖鲁时间，格林尼治标准时间（Greenwich Mean Time）

peak gust 最大阵风，尖峰阵风，[气象] 峰值阵风

statute mile 法定英里（1760 码，约合 1.6 公里）

runway visual range (RVR) 跑道视程，跑道视距，跑道能见距离

prevailing visibility [气象] 盛行能见度，主导能见度

funnel cloud 漏斗云

cloud base [气象] 云底

ceiling of cloud [气象] 云幂，云底部高度

towering cumulus (TCV) [气象] 塔状积云

altimeter setting 高度表拨定，高度拨定值，高度表拨正值

Pressure Rising Rapidly(PRESRR)
压力迅速上升

Pressure Falling Rapidly(PRESFR)
压力迅速下降

variable visibility 多变能见度，变动能见度

pilot weather reports 驾驶员气象报告

wind shear 风切变

radar weather reports 雷达气象预报

on a routine basis 常规性的，常规性地

distribution system 航行通报系统

Mean Sea Level / Mean Seal Level (MSL)
平均海平面

cloud top	[气象] 云顶，云顶，云颠
echo pattern	回声图形，回声图形
single cell	单细胞
true north	真北（以地轴北极为正北）
range set	区域集，范围集
WSR-88D weather radar	WSR-88D 气象雷达
radar summary chart	雷达综合图
grid box	网格
rain shower	阵雨

Notes

1. **International Civil Aviation Organization** 国际民用航空组织，国际民航组织

   The International Civil Aviation Organization is a specialized and funding agency of the United Nations. It changes the principles and techniques of international air navigation and fosters the planning and development of international air transport to ensure safe and orderly growth. Its headquarters is located in the Quartier International of Montreal, Quebec, Canada.

   ICAO is funded and directed by 193 national governments to support their diplomacy and cooperation in air transport as signatory states to the Chicago Convention (1944). Its core function is to maintain an administrative and expert bureaucracy (the ICAO Secretariat) supporting these diplomatic interactions, and to research new air transport policy and standardization innovations as directed and endorsed by governments through the ICAO Assembly, or by the ICAO Council which the assembly elects.

2. **Flight Service Station** 飞行服务站

   Based in Wichita, Kansas, the aviation capitol of the United States, Cessna Aircraft Company is the world's largest manufacturer of private aircraft. Cessna began its operations building small propeller-driven aircraft for the private pilot market, eventually expanding into the manufacture of corporate jets. The company has since become the leading private jet manufacturer in the industry.

   The Mission of the Cessna Aircraft Company is: to be the worldwide leader in the industry segments we serve by developing and producing safe, reliable, high-quality aircraft that represent the best value in general aviation; to provide the most comprehensive and responsive support to every Cessna customer; to produce the financial results that create value for Textron shareholders.

3. **National Weather Service** （美国）国家气象局

   The National Weather Service is an agency of the United States federal government that is tasked with providing weather forecasts, warnings of hazardous weather, and other weather-related products to organizations and the public for the purposes of protection, safety, and general information. It is a part of the National Oceanic and Atmospheric Administration branch of the Department of Commerce, and is headquartered in Silver

Spring, Maryland, within the Washington metropolitan area. The agency was known as the United States Weather Bureau from 1890 until it adopted its current name in 1970.

4. **Coordinated Universal Time** 协调世界时，协调通用时间，协调宇宙时，协道界时

Coordinated Universal Time or UTC is the primary time standard by which the world regulates clocks and time. It is within about 1 second of mean solar time at 0° longitude and is not adjusted for daylight saving time. It is effectively a successor to Greenwich Mean Time (GMT).

5. **Zulu Time** 祖鲁时间，格林尼治标准时间

Zulu Time is another word for UTC time, a system for standardizing time around the world based on time zones and the rotation of the Earth. The use of the word Zulu is connected to the NATO phonetic alphabet, in which "z" is represented by the word "Zulu".

6. **WSR-88D weather radar　WSR-88D** 气象雷达

The WSR-88D is one of the most powerful and advanced Weather Surveillance Doppler Radar in the world. Since first being built and tested in 1988, it has been installed and used operationally at over 160 locations across the United States, including Alaska and Hawaii. The WSR-88D has also been installed in Puerto Rico and several islands in the Pacific. The NWS Northern Indiana radar began warning operations on March 17th, 1998. The WSR-88D is considered by many to be the most powerful radar in the world, transmitting at 750,000 watts (an average light bulb is only 75 watts)! This power enables a beam of energy generated by the radar to travel long distances, and detect many kinds of weather phenomena. It also allows energy to continue past an initial shower or thunderstorm near the radar, thus seeing additional storms farther away. Many other radar systems do not have this kind of power, nor can they look at more than one "slice" of the atmosphere. During severe weather, the NWS WSR-88D is looking at 14 different elevations every 5 minutes, generating a radar image of each elevation. That's about 3 elevations per minute, or one radar image every 20 seconds!

Exercises

I. Answer the following questions.

1. What are the functions of aviation weather report?
2. What is METAR?
3. What do the underlined codes refer to in the following METAR?
4. What is PIREP?
5. What do the underlined codes refer to in the following RIPEP?
6. What is RAREP?
7. What do the underlined codes refer to in the following RAREP?

II. Multiple choice (one or more answers).

1. Which of the following is not aviation weather report?
   A. CAVOK.

B. METAR.

C. PIREP.

D. RAREP.

2. Through what organization is METAR standardized?

A. ICAO.

B. FSS.

C. NWS.

D. UTC.

3. What does GR refer to in METAR concerning weather phenomena?

A. Ground rain.

B. Hail.

C. Mist.

D. Dust.

4. If the amount of sky coverage is 6/8, which contraction should be used in METAR?

A. BKN.

B. CLR.

C. SCT.

D. OVC.

5. What does TM refer to in PIREPs?

A. Temperature.

B. Time.

C. Turbulence.

D. Type aircraft.

6. What does SK refer to in PIREPs?

A. Sky.

B. Cloud layers.

C. Speed.

D. Storm.

7. What kinds of echo pattern does RAREP NOT include?

A. AREA.

B. CELL.

C. LN.

D. MTS.

Text B Extensive Reading

The Most Important AVIATION WEATHER ABBREVIATIONS!

This article will be listing some of the most frequently used and the most important abbreviations regarding the weather. So, to all aspiring aviation students out there, pay attention since these abbreviations will accompany you through your whole career.

· ATIS, the Automatic Terminal Information **Service**, is the self-repeating, pre-recorded or generated voice you can listen to on a specific frequency broadcasting in the local weather airport.

· AWOS, the Automatic Weather Observing Reporting **System**, provides pilot with real-time weather information for the airport. It is about, for example, air pressure, wind speed, wind direction and air temperature, and broadcasting via radio frequency or telephone service to pilots.

· BECMG is becoming, is used when a certain change in the weather is expected within a specific **time frame**. On printouts, you will find becoming followed by a two-digit for the day of the month, then a two-digit beginning the hour, again the day of the month and a two-digit ending hour. And after that comes the expected weather condition in that time frame.

· BKN (Bravo kilo November), when isolated rays of sunlight from something called a **Jacob's Ladder**, it is likely that the ceiling is a broken. This is one of six abbreviations in this article regarding the clouds. Now, in this case, broken means that five to seven eighths of the sky are covered in clouds.

· BLZD (Bravo Lima Zulu Delta), a nasty snowstorm which is, of course, a **blizzard**. During a blizzard, there is a high risk of severe decrease in visibility and **contamination** of the wing surfaces. That's why blizzards are often accompanied by several airport closings.

· CAT (Charlie Alpha Tango), almost everyone who flew in an airplane before knows this awkward feeling in the belly when all of a sudden the airplane drops several feet. Clear Air **Turbulence** (CAT) is often the cause for that. Clear Air Turbulence are almost impossible for the pilots to detect and to predetermine, since there are no visible signs for **imminent** CATs.

· CAVOK, the abbreviation for Clouds And Visibility OK. In this case, there are no clouds below 5,000 feet above ground level and the visibility is 10 kilometers or more.

· DRZL (Delta Romeo Zulu Lima), when you are slowly getting wet, but it isn't real rain, then it's **drizzle**.

· DWPNT (Delta Whiskey Papa November Tango). The temperature at which the air would reach a hundred percent humidity is called the **dew point**. If the temperature drops below

time frame（用于某事的）一段时间

Jacob's Ladder 天梯

blizzard n. /ˈblɪzəd/ 暴风雪，雪暴
contamination n. /kənˌtæmɪˈneɪʃn/ 污染，玷污

turbulence n. /ˈtɜːrbjələns/（空气和水的）湍流，涡流，紊流
imminent adj. /ˈɪmɪnənt/ 即将发生的，临近的

drizzle n. /ˈdrɪzl/ 毛毛细雨

dew point 露点

that, water vapor will condense into clouds or fog, and will form dew or frost when coming in contact with surfaces near the ground.

· EMBDD (Echo Mike Bravo Delta Delta). Thunderstorms that are **embedded** in **cumulonimbus** cloud layers represent another **jeopardy**.

· FEW. Everyone knows it, the few clouds that occasionally disturb you while taking a sunbath on an otherwise sunny day. FEW (clouds) means one to two eighth (1/8 ~ 2/8) of the sky is covered in clouds.

· FRST (Fox Romeo Sierra Tango). As good as it looks when it coats your lawn on a cold morning, but in aviation it can be very hazardous, frost. Frost increases the weight of your aircraft, roughing the surface it contaminates, thus slowing down the airflow over the wings, resulting in flow separation and loss of lift.

· GNDFG (Gulf November Delta Fox Gulf). When visibility in mystic conditions is less than one kilometer, we talk about fog. Ground fog, however, is a phenomenon where dense fog accumulates near the ground, heavily affecting visibility.

· GSTS (Golf Sierra Tango Sierra). Something you, as a pilot, definitely want to be informed about, when coming in for landing, **wind gusts**.

· HLSTO (Hotel Lima Sierra Tango Oscar). Can little ice nuggets cause some serious damage to an airplane skin, window and engines? Of course, I'm talking about **hailstones**.

· HUREP, a natural phenomenon a pilot should definitely avoid when flying through is a hurricane. Therefore, it's important to keep track of the **hurricane** report.

· ICG. Have you ever experienced frozen wipers that won't move, or your car doors that are jammed because they are frozen? The same can happen to moveable parts on airplanes therefore it's important to deice, once icing occurs.

· IMC. Have you ever seen this? Those are view-limiting devices, or sometimes called fogged, and they are used to train flying in **Instrument Metrological Conditions**, meaning with zero visibility as you might experience in very cloudy or bad weather.

· LTG. Since an airplane is a **Faraday cage**, it protects you in case of lightning.

· METAR. You will see a lot of those in the future. The

embed v. /ɪmˈbed/ 把……牢牢地嵌入

cumulonimbus n. /ˌkjuːmələʊˈnɪmbəs/ 积雨云（常伴有雷阵雨）

jeopardy n. /ˈdʒepədi/ 处于危险境地，受到威胁

gust n. /gʌst/ 一阵强风，一阵狂风

wind gust 阵风（风速突然增大）

hailstone n. /ˈheɪlstəʊn/ 雹块，雹子

hurricane n. /ˈhʌrɪkən/（尤指西大西洋的）飓风

Instrument Metrological Conditions [气象][航] 仪表飞行气象条件

Faraday cage 法拉第笼（用于静电屏蔽）

Metrological Aerodrome Report is a standardized short message that includes a selection of different weather information which was observed at the airport it is issued for, but less information than an ATIS.

· MOD. Another level of turbulence is moderate. Moderate turbulence causes perceptible changes are no harm to the airplane passenger and crew.

· NGT (November Golf Tango). Flying, once the sun has set, has its own beauty, of course it's flying during the night.

· NSW (November Sierra Whiskey). When a former weather condition or phenomenon is about to go away, the term No Significant Weather is used, so, for example, becoming could be combined with No Significant Weather on TAP reports to signal that a previously mentioned where the condition is about to disappear.

· OVC (Oscar Victor Charlie). Have you ever heard someone say "no matter the weather, it's always sunny above the clouds"? Well, that's of course true when flying on an **overcast day**.

· PO (Papa Oscar). When you read PO (Papa Oscar), you have to pay attention to possible dust or sand whirls.

· PROB 30. It is sure that everyone knows those little percentage remarks next to the rain cloud icon on a weather forecast. Those are probabilities. And they of course occur in aviation weather forecast as well. PROB 30, for example, means that describe weather condition occurs with a probability of 30 percent. You will see that a lot.

· RMK (Romeo Mike Kilo) marks the end of the standard meter weather observations and the beginning of these remarks section. They are attached to provide the pilot with additional information. They can, for example, include lightning frequencies and types, hailstones sizes, additional cloud information and much more.

· SCT (Sierra Charlie Tango). When about half of the sky is filled with clouds, it is **scattered**. When you read this in a report, it means three to four eighths (3/8 ~ 4/8) of the sky is covered in clouds.

· SEV (Sierra Echo Victor). When large and abrupt changes in altitude make walking and the service in the cabin impossible, it is likely that you are dealing with severe turbulence. Now, there were quite a few cases that include several injuries due to severe

metrological. *adj.* /ˌmetrəˈlɒdʒɪkl/ [计量] 度量衡学的
aerodrom *n.* /ˈeərədrəʊm/ 小型飞机场

overcast *adj.* /ˌəʊvəˈkɑːst/ 阴天的，多云的，阴暗的
overcast day *n.* [气象] 阴日，阴天

scattered *adj.* /ˈskætəd/ 分散的，零散的，疏落的

turbulence in the past, so keep your seatbelts always fastened.

· SKC (Sierra Kilo Charlie). Blue sky, no clouds to be seen. The sky is clear.

· SNW (Sierra November Whiskey). The stuff that transforms everything into winter wonderland is, of course, snow.

· SQLN (Sierra Quebec Lima November). A line of heavy thunderstorms that mostly form ahead of a cold front is called a **squall line**. Now, due to large differences in pressure around this line, it is especially dangerous to fly in this area, because of strongest and occasional **downburst**s.

· STM (Sierra Tango Mike). The general abbreviation for a **storm**, expect strong winds and turbulence.

· TAF. The Terminal Aerodrome Forecast is an aviation weather forecast standardized by ICAO. Now, you're going to be seeing a lot of those in the future. Now, since it is a forecast, it doesn't provide the observed weather situation like a METAR does, therefore becoming all PROP 30 are typical examples of abbreviation used in a TAF report.

· TEMPO. Being another abbreviation you can find in a TAF report, TEMPO means that temporary weather changes are expected within a time frame attached to this term.

· THDR (Tango Hotel Delta Romeo). Now, since we've already talked about lightnings, it is inevitable to talk about this one, because no lightening without thunder.

· TS (Tango Sierra). The cause for lightnings and thunder which part try to avoid flying through is, of course, a thunderstorm.

· UWNDS (Uniform Whiskey November Delta Sierra). Now, since wind speed increases with altitude, it is important that you are able to keep track of the **upper wind**s, since wind plays an important role in the aircraft performance.

· VA (Victor Alpha). Do you remember that volcano in Iceland with the difficult name that erupted in 2010? The eruption of the Eyjafjallajökull (埃亚菲亚德拉) volcano in Iceland. Back then, the air traffic in large parts of Europe came to a complete stop because of the volcanic ash.

· VSBY (Victor Sara Bravo Yankee). Another very important point that one should never forget during a weather briefing is **visibility**.

· WS (Whiskey Sierra). Now, this phenomenon is especially dangerous for airplanes during takeoff and landing. Sudden

squall line [气象] 飑线, 阵风线
downburst
n. /ˈdaʊnbɜːst/ 下沉气流

upper wind 高空气流, [气象] 高空风

changes in wind speed and direction known as **wind shear**.

· WSR (Whiskey Sierra Romeo). To be able to accelerate or decelerate safely on a runway, an airplane has to have sufficient **traction** between its tires and the runway surface. Having wet snow on the runway is therefore an information a pilot should be provided with.

· And the last one for today is XPC (X-ray Papa Charlie). When you read this abbreviation, you can expect the certain weather condition mentioned afterwards.

Those were, in my opinion, the most common and most important weather abbreviations in aviation. If you are a flight student, I highly recommend you practice these abbreviations over and over again, because it's a little embarrassing if you don't understand what is meant.

wind shear 风切变

traction *n.* /ˈtrækʃn/ 牵引，拖拉，牵引力，拉力，车轮等对地面的）附着摩擦力

Exercises

I. Answer the following questions.

1. Why is weather abbreviation important to aviation personnel?
2. What is CAVOK?
3. What would happen if it is FEW?
4. How would GNDFG affect flight?
5. What does PROB 40 of SKC mean?

II. Multiple choice (one or more answers).

1. Which of the following abbreviations is NOT related to cloud?
   A. BKN.
   B. CAVOK.
   C. ICG.
   D. FEW.
2. Which of the following abbreviations has something to do with wind?
   A. GSTS.
   B. LTG.
   C. MOD.
   D. VA.
3. Which of the following abbreviations is NOT about storm?
   A. BLZD.
   B. CAT.
   C. EMBDD.
   D. SQLN.
   E. STM.

F. TS.

4. Which of the following abbreviations is about sand whirls?

   A. BLZD.

   B. PO.

   C. SEV.

   D. TEMPO.

5. Which of the following abbreviations is weather service provided to pilots?

   A. ATIS.

   B. AWOS.

   C. NGT.

   D. XPC.

Part 3 Translation

1. There are two types of METAR reports. The first is the routine METAR report that is transmitted every hour. The second is the aviation selected special weather report (SPECI).

   _____

   _____

2. At times, RVR, or runway visual range is reported following the prevailing visibility. RVR is the distance a pilot can see down the runway in a moving aircraft.

   _____

   _____

3. Sky condition is always reported in the sequence of amount, height, and type or indefinite ceiling/height (vertical visibility).

   _____

   _____

4. When unexpected weather conditions are encountered, pilots are encouraged to make a report to an FSS or ATC. When a pilot weather report is filed, the ATC facility or FSS will add it to the distribution system to brief other pilots and provide in-flight advisories.

   _____

   _____

5. RAREPs are especially valuable for preflight planning to help avoid areas of severe weather. However, radar only detects objects in the atmosphere that are large enough to be considered precipitation.

   _____

   _____

   _____

6. ATIS，即"自动终端情报服务"，是一种自我重复的预先录制或生成的声音，你可以用当地机场的特定频率来收听的广播。

7. AWOS，即"自动天气观测系统"，为飞行员提供机场的实时天气信息。例如，气压、风速、风向和气温，通过无线电频率或电话服务对飞行员进行广播。

8. CAVOK，是"云层及能见度良好"的缩写，在这种情况下，地面海拔约 1500 米以下没有云，能见度 10 公里以上。

9. METAR，航空例行天气报告，在实践中十分常见。它是一种标准化的短信息，其中包括一些在机场观测到的不同天气信息，但比自动终端情报服务（ATIS）提供的信息要少。

10. TAF，终端机场天气预报，是国际民航组织标准化的航空天气预报。它属于预报，不像 METAR 那样是实际观测到的天气情况。"可能性为 30%（PROP 30）"是 TAF 报告中使用缩写的典型例子。

Part 4 Supplementary Reading

The Impact of Weather on Aviation
And the Most Frequently Asked Questions
about Bad Weather and Air Travel

Whether you are a frequent flyer or occasionally take a flight to any given destination, the chances are good that you experienced a few unsettling moments during one of those flights.

With the weather getting more volatile and extreme by the year, it's time to take a serious look at the impact these meteorological conditions will have on aviation.

In this post, we examine the effect increasingly inclement weather events have on air transport, and also look at some of the most frequently asked questions about bad weather and the problems it creates in air travel.

Impact of Weather On Aviation

Even in the best weather conditions, airplanes may still experience some "inconveniences" during a flight. Turbulence, crosswinds, and air pocket are just a few normal atmospheric conditions that most passengers consider as part of a routine flight.

You can imagine how increasingly inclement weather can severely impact not just an aircraft's ability to fly but also to take off in the first place.

What is important to remember is that all forms of extreme weather can have a significant influence on flights. One usually tends to associate cold and stormy weather with canceled or delayed flights.

Although they may be the biggest culprits, they are not the only factors that may cause substantial delays in planned flights. Heatwaves, heavy rain showers, gale-force winds, and low visibility (fog) can all lead to major disruptions at airports and in the air.

Especially with weather conditions getting more extreme on an increasing basis, these events are forcing airports, air traffic controllers, and aircraft manufacturers to take action. Some of them include:

· Increased resources and more focus on fast, accurate weather forecasts at airports and air traffic control centers.

· Airports are making contingency plans for delayed or canceled flights.

· A rethink and change in airport infrastructure and runway construction to cope with worsening climate conditions.

· Aircraft manufacturers planning, building, and testing aircraft to endure the harshest weather conditions.

These are just a few of many measures already considered or implemented in the aviation industry throughout the world.

The best way to understand the different inclement weather conditions that have a significant impact on all aspects of weather is to address the questions most frequently asked about bad weather and air travel.

Read the whole article "Frequently Asked Questions about Bad Weather and Air Travel" on the website www.weather.gov.

References

1. 高培新，PEPEC 民航飞行员综合英语，北京：北京航空航天大学出版社，2016.
2. 李丹莉，戴莉新，飞行英语阅读，北京：北京航空航天大学出版社，2018.
3. 李永平，民航机务专业英语，北京：清华大学出版社，2018.
4. 刘志武，杨琼，唐宝昌，航空专业英语，北京：北京理工大学出版社，2019.
5. 申卫华，陈艾莎，飞行英语：无线电陆空通话高级教程，成都：西南交通大学出版社，2014.
6. 王晶，民航英语阅读，北京：清华大学出版社，2019.
7. 张怀兴，英汉民航词典，北京：中国民航出版社，2016.
8. 郑丽，罗军，管制英语阅读，成都：西南交通大学出版社，2018.
9. 赵迎春，陈凯军，飞机维修专业英语教程 飞机系统，北京：中国水利水电出版社，2018.
10. 赵迎春，陈凯军，飞机维修专业英语教程 飞机主要结构与部件，北京：中国水利水电出版社，2018.
11. Joe Callisto. Fly with Captain Joe. https://flywithcaptainjoe.com/. 2021.
12. D. Crane. Aviation Maintenance Technician Series – General. Aviation Supplies & Academic Inc. 2017.
13. Flight Aviation Administration. Aviation Maintenance Technician Handbook – General. Washington, DC. 2018.
14. Flight Aviation Administration. Pilot's Handbook of Aeronautical Knowledge. Washington, DC. 2016.
15. Funkid. Behind the Scene – Airport. Children's Radio UK Ltd. 2021.
16. G. Meijer. *Fundamentals of Aviation Operations*. Taylor & Francis Group. 2021.
17. T. Michmerhuizen. Aviation Mechanic Handbook. Aviation Supplies & Academic Inc. 2017.
18. News Center of COMAC. C919, The First Aircraft Independently Developed By China, Commences Its Maiden Flight In Shanghai. Commercial Aircraft Corporation of China, Ltd. (COMAC). http://english.comac.cc/news/latest/201705/16/t20170516_5220397.shtml. 2021.
19. Embry-Riddle Aeronautical University. Aircraft Systems (Video). Embry-Riddle Aeronautical University Education. https://erau.edu/. 2021.
20. A. M. Ali. "Air Traffic Control". Electrical and Electronics Engineering Benadir University.
21. G. Eason, B. Noble, and I.N. Sneddon. "On certain integrals of Lipschitz-Hankel type involving products of Bessel functions", Phil. Trans. Roy. Soc. London, vol. A247, pp. 529-551, April 1955. (references)
22. J. Clerk Maxwell. A Treatise on Electricity and Magnetism. 3rd ed., vol. 2. Oxford:

Clarendon, 1892, pp.68-73.

23. I.S. Jacobs and C.P. Bean. "Fine particles, thin films and exchange anisotropy", in *Magnetism*, vol. III, G.T. Rado and H. Suhl, Eds. New York: Academic, 1963, pp. 271-350.

24. K. Elissa, "Title of Paper if Known", unpublished.

25. R. Nicole. "Title of paper with only first word capitalized", J. Name Stand. Abbrev., in press.

26. Y. Yorozu, M. Hirano, K. Oka, and Y. Tagawa. "Electron spectroscopy studies on magneto-optical media and plastic substrate interface", IEEE Transl. J. Magn. Japan, vol. 2, pp. 740-741, August 1987 [Digests 9th Annual Conf. Magnetics Japan, p. 301, 1982].

27. M. Young. *The Technical Writer's Handbook*. Mill Valley, CA: University Science, 1989.

28. https://www.century-of-flight.net/parts-of-an-airplane-explained

29. https://cessna.txtav.com

30. https://www.cfinotebook.net/notebook/air-traffic-control/flight-service-stations

31. https://www.diamondaircraft.com/en/private-pilots/aircraft/da42/overview

32. http://en.wikipedia.org/wiki/Cessna_172

33. http://en.wikipedia.org/wiki/Piper_PA-28_Cherokee

34. http://english.comac.cc/news/latest/201705/16/t20170516_5220397.shtml

35. https://www.faa.gov

36. https://www.flightliteracy.com/ground-effect/

37. https://www.flightliteracy.com/underestimating-the-importance-of-weight-and-balance

38. http://www.free-online-private-pilot-ground-school.com/Aviation-weather-reporting.html

39. https://www.thebalancecareers.com/aircraft-weight-and-balance-terms-282771

40. https://www.thebalancecareers.com/how-do-pilots-navigate-282803

41. https://www.icao.int/Pages/default.aspx

42. https://ownyourweather.com/the-impact-of-weather-on-aviation

43. https://www.piper.com

44. https://simpleflying.com/air-traffic-controller-job/

45. https://www.weather.gov

46. https://www.timeanddate.com/time/aboututc.html

47. https://www.scienceabc.com/eyeopeners/what-is-zulu-time.html

48. https://www.weather.gov/iwx/wsr_88d

49. https://wenku.baidu.com/view/d5eccb4c33687e21af45a9ab.html